NEW VISUAL PERSPECTIVES ON
FIBONACCI NUMBERS

NEW VISUAL PERSPECTIVES ON FIBONACCI NUMBERS

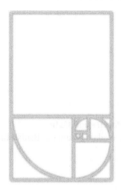

K T Atanassov

Bulgarian Academy of Sciences, Bulgaria

V Atanassova

University of Sofia, Bulgaria

A G Shannon

Warrane College, University of New South Wales, Australia

J C Turner

University of Waikato, New Zealand

World Scientific
New Jersey • London • Singapore • Hong Kong

Published by

World Scientific Publishing Co. Pte. Ltd.

P O Box 128, Farrer Road, Singapore 912805

USA office: Suite 202, 1060 Main Street, River Edge, NJ 07661

UK office: 57 Shelton Street, Covent Garden, London WC2H 9HE

British Library Cataloguing-in-Publication Data
A catalogue record for this book is available from the British Library.

ISBN 981-238-114-7
ISBN 981-238-134-1 (pbk)

Printed by FuIsland Offset Printing (S) Pte Ltd, Singapore

Introduction

There are many books now which deal with Fibonacci numbers, either explicitly or by way of examples. So why one more? What does this book do that the others do not?

Firstly, the book covers new ground from the very beginning. It is not isomorphic to any existing book. This new ground, we believe, will appeal to the research mathematician who wishes to advance the ideas still further, and to the recreational mathematician who wants to enjoy the puzzles inherent in the visual approach.

And that is the second feature which differentiates this book from others. There is a continuing emphasis on diagrams, both geometric and combinatorial, which act as a thread to tie disparate topics together – together, that is, with the unifying theme of the Fibonacci recurrence relation and various generalizations of it.

Experienced teachers know that there is great pedagogic value in getting students to draw diagrams whenever possible. These, together with the elegant identities which have always characterized Fibonacci number results, provide attractive visual perspectives. While diagrams and equations are static, the process of working through the book is a dynamic one for the reader, so that the reader begins to read in the same way as the discoverer begins to discover.

The structure of this book follows from the efforts of the four authors (both individually and collaboratively) to approach the theme from different starting points and with different styles, and so the four parts of the book can be read in any order. Furthermore, some readers will wish to focus on one or two parts only, whilst others will digest the whole book.

Like other books which deal with Fibonacci numbers, very little prior mathematical knowledge is assumed other than the rudiments of algebra and geometry, so that the book can be used as a source of enrichment material to stimulate that shrewd guessing which characterizes mathematical thinking in number theory, and which makes many parts of number theory both accessible and attractive to devotees, whether they be in high school or graduate college.

All of the mathematical results given in this book have been discovered or invented by the four authors. Some have already been published by the authors in research papers; but here they have been developed and inter-related in a new and expository manner for a wider audience. All earlier publications are cited and referenced in the Bibliographies, to direct research mathematicians to original sources.

Foreword

by A. F. Horadam

How can it be that Mathematics, being after all a product of human thought independent of experience, is so admirably adapted to the objects of reality?
— **A. Einstein**.

It has been observed that three things in life are certain: death, taxes and Fibonacci numbers. Of the first two there can be no doubt. Nor, among its devotees in the worldwide Fibonacci community, can there be little less than certainty about the third item.

Indeed, the explosive development of knowledge in the general region of Fibonacci numbers and related mathematical topics in the last few decades has been quite astonishing. This phenomenon is particularly striking when one bears in mind just what little attention had been directed to these numbers in the eight centuries since Fibonacci's lifetime, always excepting the significant contributions of Lucas in the nineteenth century.

Coupled with this expanding volume of theoretical information about Fibonacci-related matters there have been extensive ramifications in practical applications of theory to electrical networks, to computer science, and to statistics, to name only a few special growth areas. So outreaching have been the tentacles of Fibonacci-generated ideas that one ceases to be surprised when Fibonacci and Lucas entities appear seemingly as if by magic when least expected.

Several worthy texts on the basic theory of Fibonacci and Lucas numbers already provide background for those desiring a beginner's knowledge of these topics, along with more advanced details. Specialised research journals such as "The Fibonacci Quarterly", established in 1963, and "Notes on Number Theory and Discrete Mathematics", begun as recently as 1996, offer springboards for those diving into the deeper waters of the unknown.

What is distinctive about this text (and its title is most apt) is that it presents in an attractive format some new ideas, developed by recognised and experienced research workers, which readers should find compelling and stimulating. Accompanying the explanations is a wealth of striking visual images of varying complexity – geometrical figures, tree diagrams, fractals, tessellations, tilings (including polyhedra) – together with extensions for possible further research projects. A useful flow-chart suggests the connections between the number theoretic and geometric aspects of the material in the text, which actually consists of four distinct, but not discrete, components reflecting the individualistic style, tastes, and commitment of each author.

Beauty in Mathematics, it has been claimed, can be perceived, but not explained. There is much of an aesthetic nature offered here for perception, both material and physical, and we know, with Keats, that

> *A thing of beauty is a joy for ever:*
> *Its loveliness increases; ...*

Some germinal notions in the book which are ripe for exploitation and development include: the generation of pairs of sequences of inter-linked second order recurrence relations (with extensions and modifications); Fibonacci numbers and the *honeycomb plane*; the poetically designated *goldpoint geometry* associated with the golden ratio divisions of a line segment; and *tracksets*. Inherent in this last concept is the interesting investigation of the way in which group theory might have originated if Cayley had used the idea of a trackset instead of tables of group operations.

An intriguing application of goldpoint tiling geometry relates to recreational games such as chess. Indeed, there is something to be gleaned from this book by most readers.

In any wide-ranging mathematical treatise it is essential not to neglect the human aspect in research, since mathematical discoveries (e.g., zero, the irrationals, infinity, Fermat's Last Theorem, non-Euclidean geometry, Relativity theory) have originated, often with much travail and anguish, in the human mind. They did not spring, in full bloom, as the ancient Greek legend assures us that Athena sprang fully-armed from the head of the god Zeus. Readers will find some of the warmth of human association in various compartments of the material presented.

Moreover, those readers also looking for a broad and challenging outlook in a book, rather than a narrow, purely mathematical treatment (however effectively organised), will detect from time to time something of the music, the poetry, and the humour which Bertrand Russell asserted were so important to an appreciation of higher mathematics.

A suitable concluding thought emanates from Newton's famous dictum:

> *... I seem to have been only like a boy playing on the seashore, and diverting myself in now and them finding a smoother pebble or a prettier shell than ordinary, while the great ocean of truth lay all undiscovered before me.*

While much has changed since the time of Newton, there are still many glittering bright pebbles and bewitching, mysterious shells cast up by that mighty ocean (of truth) for our discovery and enduring pleasure.

A. F. Horadam

The University of New England,
Armidale, Australia

October 2001

Preface

This book presents new ideas in Fibonacci number theory and related topics, which have been discovered and developed by the authors in the past decade. In each topic, a diagrammatic or geometric approach has predominated. The illustrations themselves form an integral part of the development of the ideas, and the book in turn unravels the illustrations themselves.

There is a two-fold emphasis in the diagrams: partly to illustrate theory and examples, and partly as motivation and springboard for the development of theory. In these ways the visual illustrations are *tools of thought*, exemplifying or analogous to ideas developed by K. E. Iverson about mathematical symbols.*

The resulting *visual perspectives* comprise, in a sense, two sub-books and two sub-sub-books! That is not to say that there are four separate and unrelated monographs between the same covers. The two major parts, the number theoretic and the geometric, and the four sections are distinct, but there are many interrelations and connecting links between them.

---

*Iverson, Kenneth E. 1980: Notation as a Tool of Thought. *Communications of the Association of Computing Machinery.* Vol. 23(8), 444–465.

The following flowchart gives a simple overview of the book's structure and contents.

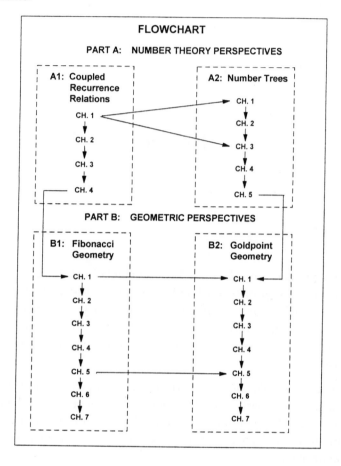

The reason for developing the book this way, rather than producing a traditional text on number theory, is to preserve the styles of the originating authors in the various parts rather than to homogenize the writing. The 'urgency' of the authors' work is thereby conserved.

The topics in the book can be entered at different points for different purposes. There are sections for various (though overlapping) audiences:

* Enrichment work for high school students,
* Background material for teacher education workshops,
* Exercises for undergraduate majors,
* Ideas for development by graduate students,
* Topics for further research by professional mathematicians,
* Enjoyment for the interested amateur (in the original sense of this word, which comes from the Latin amo: I love).

There are several common notational threads built around the sequence of Fibonacci numbers, F_n, defined by the second order homogeneous linear recurrence relation

$$F_n = F_{n-1} + F_{n-2}\,,\ n > 2,$$

with initial conditions $F_1 = F_2 = 1$.

The golden ratio is also a constant thread which links otherwise diverse topics in the various sections. It is represented here by $\alpha = (1+\sqrt{5})/2$, and it arises as the dominant solution of the auxiliary (or characteristic) polynomial equation $x^2 - x - 1 = 0$, associated with the Fibonacci recurrence relation.

In this book golden ratio gives rise in turn to new ideas relating golden means to a variety of geometric objects, such as goldpoint rings, various goldpoint fractals, and jigsaw tiles marked with goldpoints.

The geometric connections with number theory bring out some fundamental mathematical properties which are not always included in the modern school syllabus, yet they are very much part of the cultural heritage of mathematics, which, presumably, is one reason for including mathematics in a high school curriculum.

These fundamental properties are also an important component in the development of conceptual frameworks which enable mathematicians to experiment, to guess shrewdly, to test their guesses and to *see* visual perspectives in symbols, formulas and diagrams.

Thus we have presented new extensions and inventions, all with visual methods helping to drive them, in several areas of the rapidly expanding field now known as Fibonacci mathematics. We hope that readers within a wide range of mathematical abilities will find material of interest to them; and that some will be motivated sufficiently to pick up and add to our ideas.

K. T. ATANASSOV
Bulgarian Academy of Sciences, Sofia-1113, Bulgaria.

VASSIA ATANASSOVA
University of Sofia, Bulgaria.

A. G. SHANNON
Warrane College, University of New South Wales, Kensington, 1465
& KvB Institute of Technology, North Sydney, NSW, 2060, Australia.

J. C. TURNER
University of Waikato, Hamilton, New Zealand.
JCT@thenet.net.nz

May 2002

Contents

PART A. NUMBER THEORETIC PERSPECTIVES

Section 1. Coupled Recurrence Relations

Section 2. Number Trees

PART B. GEOMETRIC PERSPECTIVES

Section 1. Fibonacci Vector Geometry

Section 2. Goldpoint Geometry

PART A: NUMBER THEORETIC PERSPECTIVES

SECTION 1

COUPLED RECURRENCE RELATIONS

Krassimir Atanassov and Anthony Shannon

Coupled differential equations are well-known and arise quite naturally in applications, particularly in compartmental modelling [13]. Coupled difference equations or recurrence relations are less well known. They involve two sequences (of integers) in which the elements of one sequence are part of the generation of the other, and *vice versa*. At one level they are simple generalizations of ordinary recursive sequences, and they yield the results for those by just considering the two sequences to be identical. This can be a merely trivial confirmation of results. At another level, they provide visual patterns of relationships between the two coupled sequences which naturally leads into 'Fibonacci geometry'. In another sense again, they can be considered as the complementary picture of the intersections of linear sequences [32] for which there are many unsolved problems [25].

1

Chapter 1

Introductory Remarks by the First Author

The germ of the idea which has been unfolded with my colleagues came to me quite unexpectedly. A brief description of that event will help explain the nature of this first Section of our book.

It was a stifling hot day in the summer of 1983. I had started to work on Generalized Nets, and extension of Petri Nets, and I had searched for examples of parallel processes which are essential in Generalized Nets. After almost twenty years I can remember that day well. I was discussing my problem with colleagues at the Physical Institute of the Bulgarian Academy of Sciences, when one of them (she is an engineer) asked me: "Are there real examples of parallelism in Mathematics?"

I was not ready for such a question then. Nor do I have a good answer to it now! Nevertheless, as a young mathematician who naively thought that Mathematics readily discloses its secrets, I answered that there are obviously such examples; and I began to try to invent one on the spot.

I started by saying that the process of construction of the Fibonacci numbers is a sequential process [1,2], and began to describe the sequence and its properties. At that moment I suddenly thought of an extension of the idea, which perhaps would give an example of a parallel process. I said: "Consider two infinite sequences $\{a_n\}$ and $\{b_n\}$, which have given initial values a_1, a_2 and b_1, b_2, and which are generated for every natural number $n \geq 2$ by the coupled equations:

$$
\begin{aligned}
a_{n+2} &= b_{n+1} + b_n \\
b_{n+2} &= a_{n+1} + a_n \, .
\end{aligned}
$$

3

Is this a good example of parallelism in Mathematics?" The answer is No, although at the time it seemed to satisfy my colleagues. It did not satisfy me, however, because the process of computation of each sequence can be realized sequentially, though this is not reflected in the results.

The problem of parallelism in Mathematics, and the 'example' I had created, nagged away at me; and I continued to think about it that day. Towards evening I had invented more details, when Dimitar Sasselov, a friend of mine (now at the Harvard-Smithsonian Astrophysical Centre) came to my home. I asked him and my wife Lilija Atanassova (a fellow student and colleague at the Bulgarian Academy of Sciences) to examine some of the cases I had formulated.

This process was just an intellectual game for us. We ended our calculations, and then at that moment the following question was generated: "Why do we waste time on all this?". These results are very obvious and probably well-known. Up until this time I had not been seriously interested in the Fibonacci numbers. In the next two months I interviewed all my colleagues – mathematicians – and read books on number theory, but nowhere did I come across or find anything about such results. In the library of my Institute I found the only volumes of *The Fibonacci Quarterly* in Bulgaria and read everything which was available, but I did not find similar ideas there either. Then I decided to send my results to Professor Gerald Bergum, the Editor of the Quarterly. His answer was very encouraging.

This was the history of my first paper on the Fibonacci numbers [9]. The second one [4] is its modification. It was written some months after the positive referee's report.

In the meantime, the first paper was published and three months after this I obtained a letter from Professor Bergum with a request to referee J.-Z. Lee and J.-S. Lee's paper [22] in which almost all the results of my second paper and some other results were included. I gave a positive report on their paper and wrote to Prof. Bergum that my results were weaker and I offered him to throw them into the dustbin. However, he published first my second paper and in the next issue J.-Z. Lee and J.-S. Lee's paper. I write these words to underline the exceptional correctness of Professor Bergum. Without him I would not have worked in the area of the Fibonacci sequence at all. The next results [5; 10] were natural consequences of the first ones. I sent some of them to Professor Aldo Peretti, who published them in "Bulletin of Number Theory and Related Topics" and I am very grateful to him for this (see [6; 7]).

The essentially new direction of this research, related to these new types of the Fibonacci sequences, is related to my contacts with Professor Anthony Shannon. He wrote to me about the possibility for a graph representation of the Fibonacci sequence and I answered him with the question about the possibility for analogical representation of the new sequences. In two papers Anthony Shannon, John Turner and I showed this representation [26; 11].

In the last seven years other results related to extensions of the Fibonacci numbers were obtained. Some of them are continuations of the first one, but the others are related to new directions of Fibonacci sequence generalizations or other non-standard ideas.

Four years ago, I invited my friend the physicist Professor Peter Georgiev from Varna to research the matrix representation of the new Fibonacci sequences. When he ended his research, I helped him to finalize it. Therefore, my merit in writing of the series of (already 6) papers [14; 15; 16; 17; 18; 19] in press in "Bulletin of Number Theory and Related Topics" is in general in the beginning and end of the work, and only Peter's categorical insistence made me his co-author. With these words I would like to underline his greater credit for the matrix representation of the new Fibonacci sequences.

I must note also the research of V. Vidomenko [36], W. Spickerman, R. Joyner and R. Creech [28; 29; 30; 31], A. Shannon and R. Melham [27], S. Ando S. and M. Hayashi [1] and M. Randic, D. A. Morales and O. Araujo [24].

In this Section I would like to collect only those of my results related to the Fibonacci sequence, which are connected to ideas for new generalizations for this sequence. For this reason, the results related to their representations and applications will not be included here.

Chapter 2

The 2–Fibonacci Sequences

In this chapter we first define and study four different ways to generate pairs of integer sequences, using inter-linked second order recurrence equations.

2.1   The four 2-F-sequences

Let the arbitrary real numbers $a, b, c,$ and d be given.

There are four different ways of constructing two sequences $\{\alpha_i\}_{i=0}^{\infty}$ and $\{\beta_i\}_{i=0}^{\infty}$. We shall call them *2-Fibonacci sequences* (or 2-F-sequences). The four schemes are the following:

$$\alpha_0 = a, \ \beta_0 = b, \ \alpha_1 = c, \ \beta_1 = d$$
$$\alpha_{n+2} = \beta_{n+1} + \beta_n, \ n \geq 0 \tag{2.1}$$
$$\beta_{n+2} = \alpha_{n+1} + \alpha_n, \ n \geq 0$$

$$\alpha_0 = a, \ \beta_0 = b, \ \alpha_1 = c, \ \beta_1 = d$$
$$\alpha_{n+2} = \alpha_{n+1} + \beta_n, \ n \geq 0 \tag{2.2}$$
$$\beta_{n+2} = \beta_{n+1} + \alpha_n, \ n \geq 0$$

$$\alpha_0 = a, \ \beta_0 = b, \ \alpha_1 = c, \ \beta_1 = d$$
$$\alpha_{n+2} = \beta_{n+1} + \alpha_n, \ n \geq 0 \tag{2.3}$$
$$\beta_{n+2} = \alpha_{n+1} + \beta_n, \ n \geq 0$$

$$\alpha_0 = a, \ \beta_0 = b, \ \alpha_1 = c, \ \beta_1 = d$$
$$\alpha_{n+2} = \alpha_{n+1} + \alpha_n, \ n \geq 0 \tag{2.4}$$
$$\beta_{n+2} = \beta_{n+1} + \beta_n, \ n \geq 0$$

Graphically, the $(n+2)$-th members of the different schemes are obtained from the n-th and the $(n+1)$-th members as is shown in Figures 1–4.

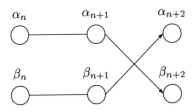

Figure 1

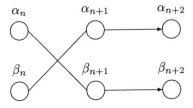

Figure 2

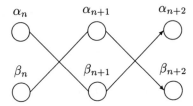

Figure 3

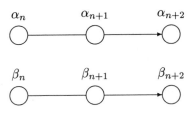

Figure 4

We shall discuss these four schemes sequentially.

2.2   The Scheme (2.1)

First, we shall study the properties of the sequences for the scheme (2.1). Clearly, if we set $a = b$ and $c = d$, then the sequences $\{\alpha_i\}_{i=0}^{\infty}$ and $\{\beta_i\}_{i=0}^{\infty}$ will coincide with each other and with the sequence $\{F_i\}_{i=0}^{\infty}$, which is called a generalized Fibonacci sequence, where

$$F_0(a, c) = a, \ F_1(a, c) = c, \ F_{n+2}(a, c) = F_{n+1}(a, c) + F_n(a, c).$$

Let $F_i = F_i(0, 1)$; $\{F_i\}_{i=0}^{\infty}$ be the ordinary Fibonacci sequence.

The first ten terms of the sequences defined in (2.1)–(2.4) are:

n	α_n	β_n
0	a	b
1	c	d
2	$b + d$	$a + c$
3	$a + c + d$	$b + c + d$
4	$a + b + 2.c + d$	$a + b + c + 2.d$
5	$a + 2.b + 2.c + 3.d$	$2.a + b + 3.c + 2.d$
6	$3.a + 2.b + 4.c + 4.d$	$2.a + 3.b + 4.c + 4.d$
7	$4.a + 4.b + 7.c + 6.d$	$4.a + 4.b + 6.c + 7.d$
8	$6.a + 7.b + 10.c + 11.d$	$7.a + 6.b + 11.c + 10.d$
9	$11.a + 10.b + 17.c + 17.d$	$10.a + 11.b + 17.c + 17.d$

A careful examination of the corresponding terms in each column leads one immediately to:

Theorem 2.1: For every integer $n \geq 0$:

$$
\begin{aligned}
(a) \quad \alpha_{3.n} + \beta_0 &= \beta_{3.n} + \alpha_0, \\
(b) \quad \alpha_{3.n+1} + \beta_1 &= \beta_{3.n+1} + \alpha_1, \\
(c) \quad \alpha_{3.n+2} + \alpha_0 + \alpha_1 &= \beta_{3.n+2} + \beta_0 + \beta_1.
\end{aligned}
$$

Proof: (a) The statement is obviously true if $n = 0$. Assume the statement is true for some integer $n \geq 1$. Then by the scheme (2.1):

$$
\begin{aligned}
\alpha_{3.n+3} + \beta_0 &= \beta_{3.n+2} + \beta_{3.n+1} + \beta_0 \ \text{(by (2.1))} \\
&= \alpha_{3.n+1} + \alpha_{3.n} + \beta_{3.n+1} + \beta_0 \ \text{(ind. hyp.)} \\
&= \alpha_{3.n+1} + \beta_{3.n} + \beta_{3.n+1} + \alpha_0 \ \text{(by (2.1))} \\
&= \alpha_{3.n+1} + \alpha_{3.n+2} + \alpha_0 \ \text{(by (2.1))} \\
&= \beta_{3.n+3} + \alpha_0 \ \text{(by (2.1))}.
\end{aligned}
$$

Hence, the statement is true for all integers $n \geq 0$. Similar proofs can be given for parts (b) and (c). $\qquad\square$

Adding the first n terms of each sequence $\{\alpha_i\}_{i=0}^{\infty}$ and $\{\beta_i\}_{i=0}^{\infty}$ yields a result similar to that obtained by adding the first n Fibonacci numbers.

Theorem 2.2: For all integer $k \geq 0$:

$$
(a) \quad \alpha_{3.k+2} = \sum_{i=0}^{3.k} \beta_i + \beta_1,
$$

$$
(b) \quad \alpha_{3.k+3} = \sum_{i=0}^{3.k+1} \alpha_i + \beta_1,
$$

$$
(c) \quad \alpha_{3.k+4} = \sum_{i=0}^{3.k+2} \beta_i + \alpha_1,
$$

$$
(d) \quad \beta_{3.k+2} = \sum_{i=0}^{3.k} \alpha_i + \alpha_1,
$$

$$
(e) \quad \beta_{3.k+3} = \sum_{i=0}^{3.k+1} \beta_i + \alpha_1,
$$

$$(f) \ \beta_{3.k+4} \ = \ \sum_{i=0}^{3.k+2} \alpha_i + \beta_1.$$

Proof: (e) if $k = 0$ the statement is obviously true, since

$$\sum_{i=0}^{1} \beta_i + \alpha_1 = \beta_0 + \beta_1 + \alpha_1 = \alpha_2 + \alpha_1 = \beta_3$$

Let us assume that (e) be true for some integer $k \geq 1$.
Then from (2.1):

$$
\begin{aligned}
\beta_{3.k+6} \ &= \ \alpha_{3.k+5} + \alpha_{3.k+4} \\
&= \ \beta_{3.k+4} + \beta_{3.k+3} + \alpha_{3.k+4} \ \text{(from (2.1))} \\
&= \ \beta_{3.k+4} + \sum_{i=0}^{3.k+1} \beta_i + \alpha_1 + \beta_{3.k+3} + \beta_{3.k+2} \\
&= \ \sum_{i=0}^{3.k+4} \beta_i + \alpha_1. \ \text{(by ind. hyp. and (2.1))}
\end{aligned}
$$

Hence, (e) is true for all integers $k \geq 0$. $\qquad\qquad\square$

Adding the first n terms with even or odd subscripts for each sequence $\{\alpha_i\}_{i=0}^{\infty}$ and $\{\beta_i\}_{i=0}^{\infty}$ we obtain more results which are similar to those obtained when one adds the first n terms of the Fibonacci sequence with even or odd subscripts. That is,

Theorem 2.3: For all integers $k \geq 0$, we have:

$$(a) \ \alpha_{6.k+5} \ = \ \sum_{i=0}^{3.k+2} \beta_{2.i} - \alpha_0 + \beta_1,$$

$$(b) \ \alpha_{6.k+6} \ = \ \sum_{i=0}^{3.k+3} \beta_{2.i-1} + \alpha_0,$$

$$(c) \ \alpha_{6.k+7} \ = \ \sum_{i=0}^{3.k+3} \beta_{2.i} - \beta_0 + \alpha_1,$$

$$(d) \ \alpha_{6.k+8} \ = \ \sum_{i=0}^{3.k+4} \beta_{2.i-1} + \beta_0,$$

$$(e) \ \alpha_{6.k+9} \ = \ \sum_{i=0}^{3.k+4} \beta_{2.i} - \beta_0 + \beta_1,$$

$$(f) \ \alpha_{6.k+9} \ = \ \sum_{i=0}^{3.k+5} \beta_{2.i-1} + \alpha_0 + \alpha_1 - \beta_1,$$

$$(g) \ \beta_{6.k+5} \ = \ \sum_{i=0}^{3.k+2} \alpha_{2.i} - \beta_0 + \alpha_1,$$

$$(h) \ \beta_{6.k+6} \ = \ \sum_{i=0}^{3.k+3} \alpha_{2.i-1} + \beta_0,$$

$$(i) \ \beta_{6.k+7} \ = \ \sum_{i=0}^{3.k+3} \alpha_{2.i} - \alpha_0 + \beta_1,$$

$$(j) \ \beta_{6.k+8} \ = \ \sum_{i=0}^{3.k+4} \alpha_{2.i-1} + \alpha_0,$$

$$(k) \ \beta_{6.k+9} \ = \ \sum_{i=0}^{3.k+4} \alpha_{2.i} - \alpha_0 + \alpha_1,$$

$$(l) \ \beta_{6.k+9} \ = \ \sum_{i=0}^{3.k+5} \alpha_{2.i-1} + \beta_0 + \beta_1 - \alpha_1.$$

Proof: (g) if $k = 0$ the statement is obviously true, since

$$\sum_{i=0}^{2} \alpha_{2.i} - \beta_0 + \alpha_1 \ = \ \alpha_0 + \alpha_2 + \alpha_4 - \beta_0 + \alpha_1$$

$$= \ 2.a + b + 3.c + 2.d = \beta_5.$$

Let us assume that (g) be true for some integer $k \geq 1$.
Then from (2.1):

$$\beta_{6.k+11} \quad = \quad \alpha_{6.k+10} + \alpha_{6.k+9}$$

$$= \quad \alpha_{6.k+10} + \beta_{6.k+9} + \alpha_0 - \beta_0 \text{ (by (2.1)a)}$$

$$= \quad \alpha_{6.k+10} + \alpha_{6.k+8} + \alpha_{6.k+7} + \alpha_0 - \beta_0$$
$$\text{(by (2.1) and ind. hyp.)}$$

$$= \quad \alpha_{6.k+10} + \alpha_{6.k+8} + \beta_{6.k+6}$$
$$+ \sum_{i=0}^{3.k+2} \alpha_{2.i} + \alpha_1 + \alpha_0 - 2.\beta_0 \text{ (Thm. 2.1 (a))}$$

$$= \quad \alpha_{6.k+10} + \alpha_{6.k+8} + \sum_{i=0}^{3.k+3} \alpha_{2.i} + \alpha_1 - \beta_0$$

$$= \quad \sum_{i=0}^{3.k+5} \alpha_{2.i} + \alpha_1 - \beta_0 \text{ (by Thm. 2.1(a))}.$$

Hence, (g) is true for all integers $k \geq 0$. A similar proof can be given for each of the remaining eleven parts of the theorem.   □

The following result is an interesting relationship which follows immediately from Theorems 2.1 and 2.2. Therefore, the proofs are omitted.

Theorem 2.4: If the integer $k \geq 0$, then

$$(a) \quad \sum_{i=0}^{3.k} (\alpha_i - \beta_i) = \alpha_0 - \beta_0,$$

$$(b) \quad \sum_{i=0}^{3.k+1} (\alpha_i - \beta_i) = \beta_2 - \alpha_2,$$

$$(c) \quad \sum_{i=0}^{3.k+2} (\alpha_i - \beta_i) = 0.$$

As one might suspect, there should be a relationship between the new sequence and the Fibonacci numbers. The next theorem establishes one of these relationships.

Theorem 2.5: If the integer $n \geq 0$, then

$$\alpha_{n+2} + \beta_{n+2} = F_{n+1}.(\alpha_0 + \beta_0) + F_{n+2}.(\alpha_1 + \beta_1).$$

Proof: The statement is obviously true if $n = 0$ and $n = 1$. Let us assume that the statement is true for all integers less than or equal to some integer $n \geq 2$. Then by (2.1):

$$\alpha_{n+3} + \beta_{n+3} = \beta_{n+2} + \beta_{n+1} + \alpha_{n+2} + \alpha_{n+1}$$

(by induction hypothesis)

$$\begin{aligned} &= \quad F_{n+1}.(\alpha_0 + \beta_0) + F_{n+2}.(\alpha_1 + \beta_1) \\ &\qquad + F_n.(\alpha_0 + \beta_0) + F_{n+1}.(\alpha_1 + \beta_1) \\ &= \quad F_{n+2}.(\alpha_0 + \beta_0) + F_{n+3}.(\alpha_1 + \beta_1). \end{aligned}$$

Hence, the statement is true for all integers $n \geq 0$. $\qquad\square$

At this point, one could continue to establish properties for the two sequences $\{\alpha_i\}_{i=0}^{\infty}$ and $\{\beta_i\}_{i=0}^{\infty}$ which are similar to those of the Fibonacci sequence. However, we have chosen another route.

Express the members of the sequences $\{\alpha_i\}_{i=0}^{\infty}$ and $\{\beta_i\}_{i=0}^{\infty}$, when $n \geq 0$, as follows:

$$\begin{aligned} \alpha_n &= \gamma_n^1.a + \gamma_n^2.b + \gamma_n^3.c + \gamma_n^4.d \\ \beta_n &= \delta_n^1.a + \delta_n^2.b + \delta_n^3.c + \delta_n^4.d \end{aligned} \qquad (2.5)$$

In this way we obtain the eight sequences $\{\gamma_i^j\}_{i=0}^{\infty}$ and $\{\delta_i^j\}_{i=0}^{\infty}$ $(j = 1, 2, 3, 4)$. The purpose of this section is to show how these eight sequences are related to each other and to the Fibonacci numbers with the major intent of finding a direct formula for calculating α_n and β_n for any n.

The following theorem establishes a relationship between these eight sequences and the Fibonacci numbers.

Theorem 2.6: For every integer $n \geq 0$:

$$(a) \; \gamma_n^1 + \delta_n^1 = F_{n-1},$$

$$(b) \ \gamma_n^2 + \delta_n^2 = F_{n-1},$$

$$(c) \ \gamma_n^3 + \delta_n^3 = F_n,$$

$$(d) \ \gamma_n^4 + \delta_n^4 = F_n.$$

Proof: (a) If $n = 0$, then:

$$\gamma_0^1 + \delta_0^1 = 1 + 0 = F_{-1}$$

and

$$\gamma_1^1 + \delta_1^1 = 0 + 0 = F_0.$$

Let us assume that the assertion is true for all integers less than or equal to some integer $n \geq 2$. Then by (2.5) and induction hypothesis

$$\begin{aligned} \gamma_{n+1}^1 + \delta_{n+1}^1 &= \delta_n^1 + \delta_{n-1}^1 + \gamma_n^1 + \gamma_{n-1}^1 \\ &= F_{n-1} + F_{n-2} = F_n, \end{aligned}$$

and, therefore, (a) is true for all integers $n \geq 0$. Similarly, one can prove parts (b), (c), and (d). $\qquad \square$

The next step is to show how the above eight sequences are related to each other.

Theorem 2.7: For every integer $k \geq 0$:

$$(a) \qquad \gamma_{3.k}^1 \ = \ \delta_{3.k}^1 + 1,$$

$$(b) \qquad \gamma_{3.k+1}^1 \ = \ \delta_{3.k+1}^1,$$

$$(c) \qquad \gamma_{3.k+2}^1 \ = \ \delta_{3.k+2}^1 - 1,$$

$$(d) \qquad \gamma_{3.k}^2 \ = \ \delta_{3.k}^2 - 1,$$

$$(e) \qquad \gamma_{3.k+1}^2 \ = \ \delta_{3.k+1}^2,$$

$$(f) \qquad \gamma_{3.k+2}^2 \ = \ \delta_{3.k+2}^2 + 1,$$

$$(g) \quad \gamma^3_{3.k} \;=\; \delta^3_{3.k},$$

$$(h) \quad \gamma^3_{3.k+1} \;=\; \delta^3_{3.k+1} + 1,$$

$$(i) \quad \gamma^3_{3.k+2} \;=\; \delta^3_{3.k+2} - 1,$$

$$(j) \quad \gamma^4_{3.k} \;=\; \delta^4_{3.k},$$

$$(k) \quad \gamma^4_{3.k+1} \;=\; \delta^4_{3.k+1} - 1,$$

$$(l) \quad \gamma^4_{3.k+2} \;=\; \delta^4_{3.k+2} + 1.$$

Proof: (j) It is obvious that (j) is true if $k = 0$, since $\gamma^4_0 = \delta^4_0 = 0$. Let us assume that the statement is true for some integer $k \geq 1$. Then by (2.1):

$$\begin{aligned}
\gamma^4_{3.k+3} &= \delta^4_{3.k+2} + \delta^4_{3.k+1} \\
&= \gamma^4_{3.k+1} + \gamma^4_{3.k} + \delta^4_{3.k+1} \text{ (by (2.1))} \\
&= \gamma^4_{3.k+1} + \delta^4_{3.k} + \delta^4_{3.k+1} \text{ (by ind. hyp.)} \\
&= \gamma^4_{3.k+1} + \gamma^4_{3.k+2} = \gamma^4_{3.k+3} \text{ (by (2.1))}
\end{aligned}$$

and the statement is proved. The remaining parts are proved in a similar way. $\qquad \square$

We now show:

Theorem 2.8: For every integer $n \geq 0$:

$$(a) \quad \gamma^1_n + \gamma^2_n = \delta^1_n + \delta^2_n,$$

$$(b) \quad \gamma^3_n + \gamma^4_n = \delta^3_n + \delta^4_n.$$

Proof: (a) This is obviously true if $n = 0$ and $n = 1$. Assume true for all integers less than or equal to some integer $n \geq 2$. Then by (2.1):

$$\gamma^1_{n+1} + \gamma^2_{n+1} \;=\; \delta^1_n + \delta^1_{n-1} + \delta^2_n + \delta^2_{n-1},$$

$$= \gamma_n^1 + \gamma_{n-1}^1 + \gamma_n^2 + \gamma_{n-1}^2 \text{ (by ind. hyp.)}$$
$$= \delta_{n+1}^1 + \delta_{n+1}^2 \text{ (by (2.1))}.$$

Similarly, one can prove part (b). $\qquad\qquad\square$

Before stating and proving our main result for this section, we need the following three theorems.

Theorem 2.9: For every integer $n \geq 0$:

$$(a)\ \delta_n^1 = \gamma_n^2,$$

$$(b)\ \delta_n^2 = \gamma_n^1,$$

$$(c)\ \delta_n^3 = \gamma_n^4,$$

$$(d)\ \delta_n^4 = \gamma_n^3,$$

$$(e)\ \gamma_n^3 = \gamma_{n+1}^2,$$

$$(f)\ \gamma_n^4 = \gamma_{n+1}^1,$$

$$(g)\ \delta_n^3 = \delta_{n+1}^2,$$

$$(h)\ \delta_n^4 = \delta_{n+1}^1.$$

Proof: (a) The statement is obviously true for $n = 0$, 1 and, 2, so assume it is true for all integers less than or equal to integer $n \geq 2$. Then by (2.1):

$$\delta_{n+1}^1 = \gamma_n^1 + \gamma_{n-1}^1$$
$$= \delta_{n-1}^1 + \delta_{n-2}^1 + \delta_{n-2}^1 + \delta_{n-3}^1 \text{ (by (2.1))}$$
$$= \gamma_{n-1}^2 + \gamma_{n-2}^2 + \gamma_{n-2}^2 + \gamma_{n-3}^2. \text{ (by ind. hyp.)}$$

And by (2.1): $\delta_n^2 + \delta_{n-1}^2 = \gamma_{n+1}^2$.

Two applications of (2.1) will complete the proof of part (a) of the theorem. The other parts are proved by similar arguments. $\qquad\qquad\square$

From Theorems 2.6 and 2.9, we have the following:

Theorem 2.10: For every integer $n \geq 0$:

$$(a) \ \gamma_n^1 + \gamma_n^2 = \delta_n^1 + \delta_n^2 = F_{n-1}$$

$$(b) \ \gamma_n^3 + \gamma_n^4 = \delta_n^3 + \delta_n^4 = F_n.$$

Finally, we have the following statement:

Theorem 2.11: For every integer $n \geq 2$:

$$(a) \ \gamma_n^1 = \gamma_{n-1}^1 + \gamma_{n-2}^1 + 3.[\tfrac{n}{3}] - n + 1,$$

$$(b) \ \gamma_n^2 = \gamma_{n-1}^2 + \gamma_{n-2}^2 - 3.[\tfrac{n}{3}] + n - 1,$$

$$(c) \ \gamma_n^1 = \gamma_n^2 + 3.[\tfrac{n}{3}] - n + 1,$$

$$(d) \ \gamma_n^3 = \gamma_{n-1}^3 + \gamma_{n-2}^3 - 3.[\tfrac{n+1}{3}] + n,$$

$$(e) \ \gamma_n^4 = \gamma_{n-1}^4 + \gamma_{n-2}^4 + 3.[\tfrac{n+1}{3}] - n,$$

$$(f) \ \gamma_n^3 = \gamma_n^4 - 3.[\tfrac{n}{3}] + n.$$

Proof: (a) The statement is obviously true if $n = 2$ or 3. Assume the statement true for all integers less than or equal to $n \geq 2$. Then by (2.1) and Theorem 2.9 (a):

$$
\begin{aligned}
\gamma_{n+1}^1 &= \delta_n^1 + \delta_{n-1}^1 = \delta_n^2 + \delta_{n-1}^2 \\[2mm]
&= \delta_{n-1}^2 + \delta_{n-2}^2 + \delta_{n-2}^2 + \delta_{n-3}^2 \ \text{(by (2.1))} \\[2mm]
&= \gamma_{n-1}^1 + \gamma_{n-2}^1 + \gamma_{n-2}^1 + \gamma_{n-3}^1 \ \text{(by Thm. 2.9 (b))}
\end{aligned}
$$

(then by the induction hypothesis:)

$$
\begin{aligned}
&= \gamma_n^1 - 3.[\tfrac{n}{3}] + n - 1 + \gamma_{n-1}^1 - 3.[\tfrac{n-1}{3}] + n - 2 \\[2mm]
&= \gamma_n^1 + \gamma_{n-1}^1 + 2.n - 3 - 3.[\tfrac{n}{3}] - 3.[\tfrac{n-1}{3}]
\end{aligned}
$$

$$= \gamma_n^1 + \gamma_{n-1}^1 + 2.n - 3 - 3.[\tfrac{n+1}{3}] - 3.n + 3$$

$$= \gamma_{n-1}^1 + \gamma_{n-2}^1 + 3.[\tfrac{n+1}{3}] - n$$

and part (a) is proved. We shall note, that it can be shown that

$$[\frac{n+1}{3}] + [\frac{n}{3}] + [\frac{n-1}{3}] = n - 1$$

for $n \geq 1$.

Similarly, one can prove the other parts. $\square$

From Theorem 2.9 (a) and Theorem 2.10 (a), we have, for $n \geq 0$,

$$\delta_{n+2}^2 = \gamma_{n+2}^1 = \frac{1}{2}.(F_{n+1} - \gamma_{n+2}^2 + \gamma_{n+1}^1 + \gamma_n^1 + 3.[\frac{n+2}{3}] - n - 1)$$

$$= \tfrac{1}{2}.(F_{n+1} - \gamma_{n+2}^2 + \delta_{n+2}^1 + 3.[\tfrac{n+2}{3}] - n - 1) \text{ (by (2.1))}$$

$$= \tfrac{1}{2}.(F_{n+1} + 3.[\tfrac{n+2}{3}] - n - 1) \text{ (by Thm. 2.9(a))}.$$

Similarly, we have

$$\gamma_{n+2}^2 = \delta_{n+2}^1 = \tfrac{1}{2}.(F_{n+1} - 3.[\tfrac{n+2}{3}] + n + 1),$$

$$\gamma_{n+2}^3 = \delta_{n+2}^4 = \tfrac{1}{2}.(F_{n+2} - 3.[\tfrac{n}{3}] + n - 1),$$

$$\gamma_{n+2}^4 = \delta_{n+2}^3 = \tfrac{1}{2}.(F_{n+2} + 3.[\tfrac{n}{3}] - n + 1).$$

Substituting these four equations into (2.5), we have:

FIRST BASIC THEOREM. If $n \geq 0$, then

$$
\begin{aligned}
\alpha_{n+2} &= \tfrac{1}{2}.((F_{n+1} + 3.[\tfrac{n+2}{3}] - n - 1).a \\
&\quad + (F_{n+1} - 3.[\tfrac{n+2}{3}] + n + 1).b \\
&\quad + (F_{n+2} - 3.[\tfrac{n}{3}] + n - 1).c \\
&\quad + (F_{n+2} + 3.[\tfrac{n}{3}] - n + 1).d) \\
&= \tfrac{1}{2}.((a + b).F_{n+1} + (c + d).F_{n+2} \\
&\quad + (3.[\tfrac{n+2}{3}] - n - 1).(a - b) + (n - 3.[\tfrac{n}{3}] - 1).(c - d), \\
\beta_{n+2} &= \tfrac{1}{2}.((F_{n+1} - 3.[\tfrac{n+2}{3}] + n + 1).a \\
&\quad + (F_{n+1} + 3.[\tfrac{n+2}{3}] - n - 1).b \\
&\quad + (F_{n+2} + 3.[\tfrac{n}{3}] - n + 1).c \\
&\quad + (F_{n+2} - 3.[\tfrac{n}{3}] + n - 1).d) \\
&= \tfrac{1}{2}.((a + b).F_{n+1} + (c + d).F_{n+2} \\
&\quad + (3.[\tfrac{n+2}{3}] - n - 1).(b - a) + (n - 3.[\tfrac{n}{3}] - 1).(d - c).
\end{aligned}
$$

2.3   The Scheme (2.2)

We shall study next the properties of the sequences for the scheme (2.2) and will conclude with a second basic theorem, as above. Since the proofs of the results below are similar to those above, we shall only list the results.

The first ten terms of the sequences defined in (2.2) are:

n	α_n	β_n
0	a	b
1	c	d
2	$b+c$	$a+d$
3	$b+c+d$	$a+c+d$
4	$a+b+c+2.d$	$a+b+2.c+d$
5	$2.a+b+2.c+3.d$	$a+3.b+3.c+2.d$
6	$3.a+2.b+4.c+4.d$	$2.a+3.b+4.c+4.d$
7	$4.a+4.b+7.c+6.d$	$4.a+4.b+6.c+7.d$
8	$6.a+7.b+11.c+10.d$	$7.a+6.b+10.c+11.d$
9	$10.a+11.b+17.c+17.d$	$11.a+10.b+17.c+17.d$

Theorem 2.12: For every integer $n \geq 0$:

(a) $\alpha_{6.n} + \beta_0 = \beta_{6.n} + \alpha_0,$

(b) $\alpha_{6.n+1} + \beta_1 = \beta_{6.n+1} + \alpha_1,$

(c) $\alpha_{6.n+2} + \alpha_0 + \beta_1 = \beta_{6.n+2} + \beta_0 + \alpha_1,$

(d) $\alpha_{6.n+3} + \alpha_0 = \beta_{6.n+3} + \beta_0,$

(e) $\alpha_{6.n+4} + \alpha_1 = \beta_{6.n+4} + \beta_1,$

(f) $\alpha_{6.n+5} + \alpha_1 + \beta_0 = \beta_{6.n+5} + \beta_1 + \alpha_0.$

Theorem 2.13: For every integer $n \geq 0$:

(a) $\alpha_{n+2} = \sum_{i=0}^{n} \beta_i + \alpha_1,$

(b) $\beta_{n+2} = \sum_{i=0}^{n} \alpha_i + \beta_1.$

Theorem 2.14: For the integer $k \geq 0$:

(a) $\displaystyle\sum_{i=0}^{6.k} (\alpha_i - \beta_i) = \alpha_0 - \beta_0,$

(b) $\displaystyle\sum_{i=0}^{6.k+1} (\alpha_i - \beta_i) = \alpha_0 - \beta_0 + \alpha_1 - \beta_1,$

(c) $\displaystyle\sum_{i=0}^{6.k+2} (\alpha_i - \beta_i) = 2.(\alpha_1 - \beta_2).$

(d) $\displaystyle\sum_{i=0}^{6.k+3} (\alpha_i - \beta_i) = -\alpha_0 + \beta_2 + 2.(\alpha_1 - \beta_1),$

(e) $\displaystyle\sum_{i=0}^{6.k+4} (\alpha_i - \beta_i) = -\alpha_0 + \beta_0 + \alpha_1 - \beta_1,$

(f) $\displaystyle\sum_{i=0}^{6.k+5} (\alpha_i - \beta_i) = 0.$

Theorem 2.15: For every integer $n \geq 0$:

$$\alpha_{n+2} + \beta_{n+2} = F_{n+1}.(\alpha_0 + \beta_0) + F_{n+2}.(\alpha_1 + \beta_1).$$

As above, we express the members of the sequences $\{\alpha_i\}_{i=0}^{\infty}$ and $\{\beta_i\}_{i=0}^{\infty}$, when $n \geq 0$ by (2.5).

It is interesting to note that the Theorems 2.6, 2.8 and 2.9 with identical forms are valid here.

Let ψ be the integer function defined for every $k \geq 0$ by:

r	$\psi(6.k + r)$
0	1
1	0
2	-1
3	-1
4	0
5	1

Obviously, for every $n \geq 0$, $\psi(n+3) = -\psi(n)$.

Using the definition of the function ψ, the following are easily proved by induction:

Theorem 2.16: For every integer $n \geq 0$:

$$(a) \; \gamma_n^1 = \delta_n^1 + \psi(n),$$

$$(b) \; \gamma_n^2 = \delta_n^2 + \psi(n+3),$$

$$(c) \; \gamma_n^3 = \delta_n^3 + \psi(n+4),$$

$$(d) \; \gamma_n^4 = \delta_n^4 + \psi(n+1).$$

Theorem 2.17: For every integer $n \geq 0$:

$$(a) \quad \gamma_{n+2}^1 = \gamma_{n+1}^1 + \gamma_n^1 + \psi(n+3),$$

$$(b) \quad \gamma_{n+2}^2 = \gamma_{n+1}^2 + \gamma_n^2 + \psi(n),$$

$$(c) \quad \gamma_n^1 = \gamma_n^2 + \psi(n),$$

$$(d) \quad \gamma_{n+2}^3 = \gamma_{n+1}^3 + \gamma_n^3 + \psi(n+4),$$

$$(e) \quad \gamma_{n+2}^4 = \gamma_{n+1}^4 + \gamma_n^4 + \psi(n+1),$$

$$(f) \quad \gamma_n^3 = \gamma_n^4 + \psi(n+4).$$

From the above theorems, we obtain the equations:

$$\gamma_n^1 = \delta_n^2 = \frac{1}{2} \cdot (F_{n-1} + \psi(n)),$$

$$\gamma_n^2 = \delta_n^1 = \frac{1}{2} \cdot (F_{n-1} + \psi(n+3)),$$

$$\gamma_n^3 = \delta_n^4 = \frac{1}{2} \cdot (F_{n-1} + \psi(n+4)),$$

$$\gamma_n^4 = \delta_n^3 = \frac{1}{2} \cdot (F_{n-1} + \psi(n+1)).$$

We may now state our second basic theorem.

SECOND BASIC THEOREM. If $n \geq 0$, then

$$\alpha_n = \tfrac{1}{2}.((F_{n-1} + \psi(n)).a + (F_{n-1}\psi(n+3).b$$

$$+ (F_n + \psi(n+4)).c + (F_n + \psi(n+1)).d)$$

$$= \tfrac{1}{2}.((a+b).F_{n-1} + (c+d).F_n$$

$$+ \psi(n).a + \psi(n+3).b + \psi(n+4).c + \psi(n+1),$$

$$\beta_n = \tfrac{1}{2}.((F_{n-1} + \psi(n+3)).a + (F_{n-1}\psi(n).b$$

$$+ (F_n + \psi(n+1)).c + (F_n + \psi(n+4)).d)$$

$$= \tfrac{1}{2}.((a+b).F_{n-1} + (c+d).F_n$$

$$+ \psi(n+3).a + \psi(n).b + \psi(n+1).c + \psi(n+4).$$

The sequences (2.4) are actually two independent Fibonacci sequences of the form $\{F_i(a,c)\}_{i=0}^{\infty}$ and $\{F_i(b,d)\}_{i=0}^{\infty}$. It is easily seen that the sequences (2.3) can be expressed through the sequences $\{F_i(a,d)\}_{i=0}^{\infty}$ and $\{F_i(b,c)\}_{i=0}^{\infty}$, for $n \geq 1$, thus:

$$\alpha_{2.n} = F_{2.n}(a,d),$$

$$\alpha_{2.n+1} = F_{2.n+1}(b,c),$$

$$\beta_{2.n} = F_{2.n}(b,c),$$

$$\beta_{2.n+1} = F_{2.n+1}(a,d),$$

On the basis of what has been done above, one could be led to generalize and examine sequences of the following types:

$$\alpha_0 \;=\; a, \; \beta_0 = b, \; \alpha_1 = c, \; \beta_1 = d$$

$$\alpha_{n+2} \;=\; p.\beta_{n+1} + q.\beta_n, \; n \geq 0$$

$$\beta_{n+2} \;=\; r.\alpha_{n+1} + s.\alpha_n, \; n \geq 0$$

$$\alpha_0 \;=\; a, \; \beta_0 = b, \; \alpha_1 = c, \; \beta_1 = d$$

$$\alpha_{n+2} \;=\; p.\alpha_{n+1} + q.\beta_n, \; n \geq 0$$

$$\beta_{n+2} \;=\; r.\beta_{n+1} + s.\alpha_n, \; n \geq 0$$

for the fixed real numbers p, q, r, and s.

This problem was formulated in 1986, but up to this moment it is open.

Finally, we shall describe one more modification of the above schemes, and outline the solutions. The new schemes have the following forms:

I type (trivial):
$\alpha_0 = a, \; \beta_0 = b, \; \alpha_1 = c, \; \beta_1 = d$
$\alpha_{n+2} = \alpha_{n+1} - \alpha_n, \; n \geq 0$
$\beta_{n+2} = \beta_{n+1} - \beta_n, \; n \geq 0$

II type (trivial):
$\alpha_0 = a, \; \beta_0 = b, \; \alpha_1 = c, \; \beta_1 = d$
$\alpha_{n+2} = \beta_{n+1} - \alpha_n, \; n \geq 0$
$\beta_{n+2} = \alpha_{n+1} - \beta_n, \; n \geq 0$

III type:
$\alpha_0 = a, \; \beta_0 = b, \; \alpha_1 = c, \; \beta_1 = d$
$\alpha_{n+2} = \alpha_{n+1} - \beta_n, \; n \geq 0$
$\beta_{n+2} = \beta_{n+1} - \alpha_n, \; n \geq 0$

IV type:
$\alpha_0 = a, \; \beta_0 = b, \; \alpha_1 = c, \; \beta_1 = d$
$\alpha_{n+2} = \beta_{n+1} - \beta_n, \; n \geq 0$
$\beta_{n+2} = \alpha_{n+1} - \alpha_n, \; n \geq 0$

For the first case (I type) we obtain directly, that:

$$\alpha_n = \begin{cases} a, & \text{if } n \equiv 0 \pmod 6 \\ c, & \text{if } n \equiv 1 \pmod 6 \\ c - a, & \text{if } n \equiv 2 \pmod 6 \\ -a, & \text{if } n \equiv 3 \pmod 6 \\ -c, & \text{if } n \equiv 4 \pmod 6 \\ a - c, & \text{if } n \equiv 5 \pmod 6 \end{cases}$$

$$\beta_n = \begin{cases} b, & \text{if } n \equiv 0 \pmod 6 \\ d, & \text{if } n \equiv 1 \pmod 6 \\ d - b, & \text{if } n \equiv 2 \pmod 6 \\ -b, & \text{if } n \equiv 3 \pmod 6 \\ -d, & \text{if } n \equiv 4 \pmod 6 \\ b - d, & \text{if } n \equiv 5 \pmod 6 \end{cases}$$

and for the second case (II type), we find that:

$$\alpha_n = \begin{cases} a, & \text{if } n \equiv 0 \pmod 6 \\ c, & \text{if } n \equiv 1 \pmod 6 \\ d - a, & \text{if } n \equiv 2 \pmod 6 \\ -b, & \text{if } n \equiv 3 \pmod 6 \\ -d, & \text{if } n \equiv 4 \pmod 6 \\ b - c, & \text{if } n \equiv 5 \pmod 6 \end{cases}$$

$$\beta_n = \begin{cases} b, & \text{if } n \equiv 0 \pmod 6 \\ d, & \text{if } n \equiv 1 \pmod 6 \\ c - b, & \text{if } n \equiv 2 \pmod 6 \\ -a, & \text{if } n \equiv 3 \pmod 6 \\ -c, & \text{if } n \equiv 4 \pmod 6 \\ a - d, & \text{if } n \equiv 5 \pmod 6 \end{cases}$$

When the other two cases are valid, the following two theorems can be proved analagously to the respective ones given above for the first schemes.

Theorem 2.18: For every integer $n \geq 0$, for the third scheme:

$$\alpha_n = \tfrac{1}{2}.((F_{n-1} + \psi(n)).a - (F_{n-1} + \psi(n+3)).b$$
$$+ (F_n + \psi(n+4)).c - (F_n + \psi(n+1)).d),$$

$$\beta_n = \tfrac{1}{2}.(-(F_{n-1} + \psi(n+3)).a + (F_{n-1} + \psi(n).b$$
$$- (F_n + \psi(n+1)).c + (F_n + \psi(n+4)).d).$$

Theorem 2.19: For every integer $n \geq 0$, for the fourth scheme:

$$\alpha_n = \tfrac{1}{2}.((-1)^n.(F_{n-1} + 3.[\tfrac{n+2}{3}] - n + 1).a$$
$$+ (-1)^{n+1}.(F_{n-1} - 3.[\tfrac{n+2}{3}] + n - 1).b$$
$$+ (-1)^{n+1}.(F_n - 3.[\tfrac{n-2}{3}] + n - 3).c$$
$$+ (-1)^n.(F_n - 3.[\tfrac{n-2}{3}] - n + 3).d),$$

$$\beta_n = \tfrac{1}{2}.((-1)^{n+1}.(F_{n-1} - 3.[\tfrac{n}{3}] + n - 1).a$$
$$+ (-1)^n.(F_{n-1} + 3.[\tfrac{n}{3}] - n - 1).b$$
$$+ (-1)^n.(F_n + 3.[\tfrac{n-2}{3}] - n + 3).c$$
$$+ (-1)^{n+1}.(F_n - 3.[\tfrac{n-2}{3}] + n - 3).d).$$

Chapter 3

Extensions of the Concepts of 2–Fibonacci Sequences

A new direction for generalizing the Fibonacci sequence was described in Chapter 2. Here, we shall continue that direction of research.

Let $C_1, C_2, ..., C_6$ be fixed real numbers. Using C_1 to C_6, we shall construct new schemes which are of the Fibonacci type and called 3-F-sequences. Let for $n \geq 0$

$$a_0 = C_1, \ b_0 = C_2, \ c_0 = C_3, \ a_1 = C_4, \ b_1 = C_5, \ c_1 = C_6$$

$$a_{n+2} = x^1_{n+1} + y^1_n$$

$$b_{n+2} = x^2_{n+1} + y^2_n$$

$$c_{n+2} = x^3_{n+1} + y^3_n$$

where $< x^1_{n+1}, x^2_{n+1}, x^3_{n+1} >$ is any permutation of $< a_{n+1}, b_{n+1}, c_{n+1} >$ and $< x^1_n, x^2_n, x^3_n >$ is any permutation of $< a_n, b_n, c_n >$.

The number of different schemes is obviously 36.

In [22], the specific scheme $(n \geq 0)$

$$a_0 = C_1, \ b_0 = C_2, \ c_0 = C_3, \ a_1 = C_4, \ b_1 = C_5, \ c_1 = C_6$$

$$a_{n+2} = b_{n+1} + c_n$$

$$b_{n+2} = c_{n+1} + a_n$$

$$c_{n+2} = a_{n+1} + b_n$$

is discussed in detail. For the sake of brevity, we devise the following representation for this scheme:

$$S = \begin{pmatrix} a & b & c \\ b & c & a \\ c & a & b \end{pmatrix} \tag{3.1}$$

Note that we have merely eliminated the subscripts and the equal and plus symbols so that our notation is similar to that used in representing a system of linear equations in matrix form. Using this notation, it is important to remember that the elements in their first column are always in the same order while the elements in the order column can be permuted within that column. Every elememt a, b and c must be used in each column.

We now define an operation called substitution over these 3-F-sequences and adopt the notation $[p, q]S$, where $p, q \in \{a, b, c\}$, $p \neq q$. Applying the operation to S merely interchanges all occurrences of p and q in each column. For example, using (3.1), we have

$$[a, c]S = \begin{pmatrix} c & b & a \\ b & a & c \\ a & c & b \end{pmatrix} \tag{3.2}$$

Note that in the result we do not maintain the order of the elements in the first column. To maintain this order we interchange the first and last rows of (3.2) to obtain:

$$S' = \begin{pmatrix} a & c & b \\ b & a & c \\ c & b & a \end{pmatrix} \tag{3.3}$$

which corresponds to the scheme $(n \geq 0)$

$$a_0 = C_1', \ b_0 = C_2', \ c_0 = C_3', \ a_1 = C_4', \ b_1 = C_5', \ C_1' = C_6'$$

$$a_{n+2} = c_{n+1} + b_n$$

$$b_{n+2} = a_{n+1} + c_n$$

$$c_{n+2} = b_{n+1} + a_n$$

where $C_1', C_2', ..., C_6'$ are real numbers.

We shall say that the two schemes S and S' are equivalent under the operation of substitution and denote this by $S \leftrightarrow S'$.

It is now obvious that for any two 3-F-sequences S and S', if $[p, q]S \leftrightarrow S'$, then $[p, q]S' \leftrightarrow S$. To investigate the concept of equivalence to a deeper extent, it is necessary to list all 36 schemes:

$$S_1 = \begin{pmatrix} a & a & a \\ b & b & b \\ c & c & c \end{pmatrix} \quad S_2 = \begin{pmatrix} a & a & a \\ b & b & c \\ c & c & b \end{pmatrix} \quad S_3 = \begin{pmatrix} a & a & a \\ b & c & b \\ c & b & c \end{pmatrix}$$

$$S_4 = \begin{pmatrix} a & a & a \\ b & c & c \\ c & b & b \end{pmatrix} \quad S_5 = \begin{pmatrix} a & a & b \\ b & b & a \\ c & c & c \end{pmatrix} \quad S_6 = \begin{pmatrix} a & a & b \\ b & b & c \\ c & c & a \end{pmatrix}$$

$$S_7 = \begin{pmatrix} a & a & b \\ b & c & a \\ c & b & c \end{pmatrix} \quad S_8 = \begin{pmatrix} a & a & b \\ b & c & c \\ c & b & a \end{pmatrix} \quad S_9 = \begin{pmatrix} a & a & c \\ b & b & a \\ c & c & b \end{pmatrix}$$

$$S_{10} = \begin{pmatrix} a & a & c \\ b & b & b \\ c & c & a \end{pmatrix} \quad S_{11} = \begin{pmatrix} a & a & c \\ b & c & a \\ c & b & b \end{pmatrix} \quad S_{12} = \begin{pmatrix} a & a & c \\ b & c & b \\ c & b & a \end{pmatrix}$$

$$S_{13} = \begin{pmatrix} a & b & a \\ b & a & b \\ c & c & c \end{pmatrix} \quad S_{14} = \begin{pmatrix} a & b & a \\ b & a & c \\ c & c & b \end{pmatrix} \quad S_{15} = \begin{pmatrix} a & b & a \\ b & c & b \\ c & a & c \end{pmatrix}$$

$$S_{16} = \begin{pmatrix} a & b & a \\ b & c & c \\ c & a & b \end{pmatrix} \quad S_{17} = \begin{pmatrix} a & b & b \\ b & a & a \\ c & c & c \end{pmatrix} \quad S_{18} = \begin{pmatrix} a & b & b \\ b & a & c \\ c & c & a \end{pmatrix}$$

$$S_{19} = \begin{pmatrix} a & b & b \\ b & c & a \\ c & a & c \end{pmatrix} \quad S_{20} = \begin{pmatrix} a & b & b \\ b & c & c \\ c & a & a \end{pmatrix} \quad S_{21} = \begin{pmatrix} a & b & c \\ b & a & a \\ c & c & b \end{pmatrix}$$

$$S_{22} = \begin{pmatrix} a & b & c \\ b & a & b \\ c & c & a \end{pmatrix} \qquad S_{23} = \begin{pmatrix} a & b & c \\ b & c & a \\ c & a & b \end{pmatrix} \qquad S_{24} = \begin{pmatrix} a & b & c \\ b & c & b \\ c & a & a \end{pmatrix}$$

$$S_{25} = \begin{pmatrix} a & c & a \\ b & a & b \\ c & b & c \end{pmatrix} \qquad S_{26} = \begin{pmatrix} a & c & a \\ b & a & c \\ c & b & b \end{pmatrix} \qquad S_{27} = \begin{pmatrix} a & c & a \\ b & b & b \\ c & a & c \end{pmatrix}$$

$$S_{28} = \begin{pmatrix} a & c & a \\ b & b & c \\ c & a & b \end{pmatrix} \qquad S_{29} = \begin{pmatrix} a & c & b \\ b & a & a \\ c & b & c \end{pmatrix} \qquad S_{30} = \begin{pmatrix} a & c & b \\ b & a & c \\ c & b & a \end{pmatrix}$$

$$S_{31} = \begin{pmatrix} a & c & b \\ b & b & a \\ c & a & c \end{pmatrix} \qquad S_{32} = \begin{pmatrix} a & c & b \\ b & b & c \\ c & a & a \end{pmatrix} \qquad S_{33} = \begin{pmatrix} a & c & c \\ b & a & a \\ c & b & c \end{pmatrix}$$

$$S_{34} = \begin{pmatrix} a & c & c \\ b & a & b \\ c & b & a \end{pmatrix} \qquad S_{35} = \begin{pmatrix} a & c & c \\ b & b & a \\ c & a & b \end{pmatrix} \qquad S_{36} = \begin{pmatrix} a & c & c \\ b & b & b \\ c & a & a \end{pmatrix}$$

Note that $S = S_{23}$ and $S' = S_{30}$, so that $S_{23} \leftrightarrow S_{30}$.

We say that a 3-F-sequence S is trivial if at least one of the resulting sequences is a Fibonacci sequence. Otherwise, S is said to be an essential generalization of the Fibonacci sequence.

Observe that there are ten trivial 3-F-sequences. They are S_1, S_2, S_3, S_4, S_5, S_{10}, S_{13}, S_{17} , S_{27} and S_{36}.

These 10 schemes are easy to detect since they have at least one row all with the same letter. Furthermore; for these schemes one of the three possible substitutions returns the scheme itself. For example,

$$[b, c]S_i \leftrightarrow S_i, \ i = 1, 2, 3, 4$$

$$[a, b]S_i \leftrightarrow S_i, \ i = 1, 5, 13, 17$$

$$[a, c]S_i \leftrightarrow S_i, \ i = 1, 10, 27, 36.$$

The twenty-six remaining schemes are essential generalizations of the Fibonacci sequence. For eight of these schemes, the result is independent of the substitution made. That is,

$$[p,q]S_6 \leftrightarrow S_9$$

$$[p,q]S_{15} \leftrightarrow S_{25}$$

$$[p,q]S_{20} \leftrightarrow S_{33}$$

$$[p,q]S_{23} \leftrightarrow S_{30}$$

for all $p, q \in \{a, b, c\}$. This means the substitution operation for these schemes is cyclic of length 2.

For the other eighteen essential generalizations of the Fibonacci sequence schemes, all three possible substitutions generate three different schemes. For example,

$$[b,c]S_7 \leftrightarrow S_{12} \quad [b,c]S_8 \leftrightarrow S_{11} \quad [b,c]S_{16} \leftrightarrow S_{26}$$

$$[a,c]S_7 \leftrightarrow S_{14} \quad [a,c]S_8 \leftrightarrow S_{21} \quad [a,c]S_{16} \leftrightarrow S_{29}$$

$$[a,b]S_7 \leftrightarrow S_{31} \quad [a,b]S_8 \leftrightarrow S_{35} \quad [a,b]S_{16} \leftrightarrow S_{34}$$

All of the substitutions associated with the remaining eighteen schemes and their results are conveniently illustrated by the following three figures. That is, these pictures determine all possible cycles.

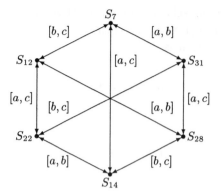

Figure 1

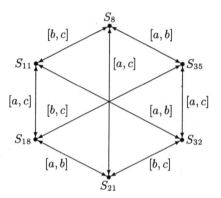

Figure 2

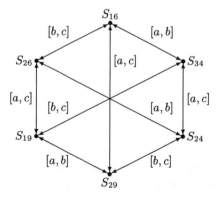

Figure 3

For example,

$$[a, b]S_{29} \leftrightarrow S_{19}$$

and

$$[a, b]([b, c]([a, c]S_{24})) \leftrightarrow S_{29}.$$

Note that the figures tell us that many of the schemes are independent. That is, S_{18} and S_{24} are independent. In fact, S_{18} is related only to the six schemes listed in Fig. 3. Similar results can be found for the other schemes.

The closed form equation of the members for all three sequences of the scheme S_{23} is given in [22].

For every one of the above groups we shall give the recurrence relations of its members.

Let everywhere

$$a_0 = C_1, \ b_0 = C_2, \ c_0 = C_3, \ a_1 = C_4, \ b_1 = C_5, \ c_1 = C_6$$

and $n \geq 0$ be a natural number, where $C_1, C_2, ..., C_6$ are given constants and 'x' is one of the symbols 'a', 'b' and 'c'.

Group I contains the schemes S_6 and S_9, where:

$$S_6 : \begin{cases} a_{n+2} = a_{n+1} + b_n \\ b_{n+2} = b_{n+1} + c_n \\ c_{n+2} = c_{n+1} + a_n \end{cases}$$

The recurrence relation for this scheme is:

$$x_{n+6} = 3.x_{n+5} - 3.x_{n+4} + x_{n+3} + x_n,$$

i.e.

$$a_{n+6} = 3.a_{n+5} - 3.a_{n+4} + a_{n+3} + a_n,$$

$$b_{n+6} = 3.b_{n+5} - 3.b_{n+4} + b_{n+3} + b_n,$$

$$c_{n+6} = 3.c_{n+5} - 3.c_{n+4} + c_{n+3} + c_n.$$

Group II contains the schemes S_{15} and S_{25} , where:

$$S_{15} : \begin{cases} a_{n+2} = b_{n+1} + a_n \\ b_{n+2} = c_{n+1} + b_n \\ c_{n+2} = a_{n+1} + c_n \end{cases}$$

The recurrence relation for this scheme is:

$$x_{n+6} = 3.x_{n+4} + x_{n+3} - 3.x_{n+2} + x_n.$$

Group III contains the schemes S_{20} and S_{33} , where:

$$S_{20} : \begin{cases} a_{n+2} = b_{n+1} + b_n \\ b_{n+2} = c_{n+1} + c_n \\ c_{n+2} = a_{n+1} + a_n \end{cases}$$

The recurrence relation for this scheme is:

$$x_{n+6} = x_{n+3} + 3.x_{n+2} + 3.x_{n+1} + x_n.$$

Group IV contains the schemes S_{23} and S_{33} , where:

$$S_{23} : \begin{cases} a_{n+2} = b_{n+1} + c_n \\ b_{n+2} = c_{n+1} + a_n \\ c_{n+2} = a_{n+1} + b_n \end{cases}$$

The recurrence relation for this scheme is:

$$x_{n+6} = 4.x_{n+3} + x_n.$$

Group V contains the schemes $S_7, S_{12}, S_{14}, S_{22}, S_{28}$ and S_{31}, where:

$$S_7 : \begin{cases} a_{n+2} = a_{n+1} + b_n \\ b_{n+2} = c_{n+1} + a_n \\ c_{n+2} = b_{n+1} + c_n \end{cases}$$

The recurrence relation for this scheme is:

$$x_{n+6} = x_{n+5} + 2.x_{n+4} - 2.x_{n+3} + x_{n+2} - x_n.$$

Group VI contains the schemes $S_8, S_{11}, S_{18}, S_{21}, S_{32}$ and S_{35}, where:

$$S_8 : \begin{cases} a_{n+2} = a_{n+1} + b_n \\ b_{n+2} = c_{n+1} + c_n \\ c_{n+2} = b_{n+1} + a_n \end{cases}$$

The recurrence relation for this scheme is:

$$x_{n+6} = x_{n+5} + x_{n+4} - x_{n+2} + x_{n+1} + x_n.$$

Group VII contains the schemes $S_{16}, S_{19}, S_{24}, S_{26}, S_{29}$ and S_{34}, where:

$$S_{16} : \begin{cases} a_{n+2} = b_{n+1} + a_n \\ b_{n+2} = c_{n+1} + c_n \\ c_{n+2} = a_{n+1} + b_n \end{cases}$$

The recurrence relation for this scheme is:

$$x_{n+6} = x_{n+4} + 2.x_{n+3} + 2.x_{n+2} - x_{n+1} - x_n.$$

Following the above idea and the idea from [22], we can construct 8 different schemes of generalized Tribonacci sequences in the case of two sequences. We introduce their recurrence relations below.

Let 'x' be one of the symbols 'a' and 'b'.

The different schemes are as following:

$$T_1 : \begin{cases} a_{n+3} = a_{n+2} + a_{n+1} + a_n \\ b_{n+3} = b_{n+2} + b_{n+1} + b_n \end{cases}$$

$$T_2 : \begin{cases} a_{n+3} = a_{n+2} + a_{n+1} + b_n \\ b_{n+3} = b_{n+2} + b_{n+1} + a_n \end{cases}$$

$$T_3 : \begin{cases} a_{n+3} = a_{n+2} + b_{n+1} + a_n \\ b_{n+3} = b_{n+2} + a_{n+1} + b_n \end{cases}$$

$$T_4 : \begin{cases} a_{n+3} = a_{n+2} + b_{n+1} + b_n \\ b_{n+3} = b_{n+2} + a_{n+1} + a_n \end{cases}$$

$$T_5 : \begin{cases} a_{n+3} = b_{n+2} + a_{n+1} + a_n \\ b_{n+3} = a_{n+2} + b_{n+1} + b_n \end{cases}$$

$$T_6 : \begin{cases} a_{n+3} = b_{n+2} + a_{n+1} + b_n \\ b_{n+3} = a_{n+2} + b_{n+1} + a_n \end{cases}$$

$$T_7 : \begin{cases} a_{n+3} = b_{n+2} + b_{n+1} + a_n \\ b_{n+3} = a_{n+2} + a_{n+1} + b_n \end{cases}$$

$$T_8 : \begin{cases} a_{n+3} = b_{n+2} + b_{n+1} + b_n \\ b_{n+3} = a_{n+2} + a_{n+1} + a_n \end{cases}$$

The first scheme is trivial. All of the others are nontrivial; they have the following recurrent formulas for $n \geq 0$:

$$
\begin{aligned}
\text{for } T_2: \quad x_{n+6} &= 2.x_{n+5} + x_{n+4} - 2.x_{n+3} - x_{n+2} + x_n, \\
\text{for } T_3: \quad x_{n+6} &= 2.x_{n+5} - x_{n+4} + 2.x_{n+3} - x_{n+2} - x_n, \\
\text{for } T_4: \quad x_{n+6} &= 2.x_{n+5} - x_{n+4} + x_{n+2} + x_{n+1} + x_n, \\
\text{for } T_5: \quad x_{n+6} &= 3.x_{n+4} + 2.x_{n+3} - x_{n+2} - 2.x_{n+1} - x_n, \\
\text{for } T_6: \quad x_{n+6} &= 3.x_{n+4} + x_{n+2} + x_n, \\
\text{for } T_7: \quad x_{n+6} &= x_{n+4} + 4.xn + 3 + x_{n+2} - x_n, \\
\text{for } T_8: \quad x_{n+6} &= x_{n+4} + 2.x_{n+3} + 3.x_{n+2} + 2.x_{n+1} + x_n.
\end{aligned}
$$

The proofs for these results can be shown by induction, using methods similar to those in Chapter 2.

An open problem is the construction of an explicit formula for each of the schemes given above.

Chapter 4

Other Ideas for Modification of the Fibonacci Sequence

In this chapter we shall describe some new ideas which are, in general, generated by the first one. After that, we can think about the question: How will the 'Fibonacci numbers world' look (because, the formidable quantity and quality of researches in this mathematical area generate one world!) when modified in the sense of the parallel sequences? Can we show essential examples of parallel processes in this world?

4.1 Remark on a new direction for a generalization of the Fibonacci sequence

Combining Peter Hope's idea from [21] with the above ideas for a generalization of the Fibonacci sequence, we can introduce a new direction for a generalization of the Fibonacci sequence (see also [8]). At the moment, all generalizations of this sequence are 'linear'. The sequence proposed here has a 'multiplicative' form. The analogue of the standard Fibonacci sequence in this form will be:

$$x_0 = a, x_1 = b, x_{n+2} = x_{n+1}.x_n (n \geq 0),$$

where a and b are real numbers. These are types of Fibonacci *words* (see [34]). Directly it can be seen, that for $n \geq 1$:

$$x_n = a^{f_{n-1}}.b^{f_n}.$$

In the case of two (or more) sequences, by analogy with the development in Chapter 2, we shall define the following four schemes:

$$\alpha_0 \;=\; a,\; \beta_0 = b,\; \alpha_1 = c,\; \beta_1 = d$$
$$\alpha_{n+2} \;=\; \beta_{n+1}.\beta_n,\; n \geq 0$$
$$\beta_{n+2} \;=\; \alpha_{n+1}.\alpha_n,\; n \geq 0$$

$$\alpha_0 \;=\; a,\; \beta_0 = b,\; \alpha_1 = c,\; \beta_1 = d$$
$$\alpha_{n+2} \;=\; \alpha_{n+1}.\beta_n,\; n \geq 0$$
$$\beta_{n+2} \;=\; \beta_{n+1}.\alpha_n,\; n \geq 0$$

$$\alpha_0 \;=\; a,\; \beta_0 = b,\; \alpha_1 = c,\; \beta_1 = d$$
$$\alpha_{n+2} \;=\; \beta_{n+1}.\alpha_n,\; n \geq 0$$
$$\beta_{n+2} \;=\; \alpha_{n+1}.\beta_n,\; n \geq 0$$

$$\alpha_0 \;=\; a,\; \beta_0 = b,\; \alpha_1 = c,\; \beta_1 = d$$
$$\alpha_{n+2} \;=\; \alpha_{n+1}.\alpha_n,\; n \geq 0$$
$$\beta_{n+2} \;=\; \beta_{n+1}.\beta_n,\; n \geq 0$$

Obviously, these schemes are analogous to the 'additive' schemes. Thus we can call them 'multiplicative' schemes.

The n-th terms of these schemes are determined, e.g., as constructed in Chapter 2. We shall give the formulas for the $(n+2)$-th terms $(n \geq 0)$, using the above notation. These terms are as follows:

- Scheme I:

$$\alpha_{n+2} \;=\; a^{\frac{1}{2}.(F_{n+1}+3.[\frac{n+2}{3}]-n-1)}.b^{\frac{1}{2}.(F_{n+1}-3.[\frac{n+2}{3}]+n+1)}.c^{\frac{1}{2}.(F_{n+2}-3.[\frac{n}{3}]+n-1)}$$
$$.d^{\frac{1}{2}.(F_{n+2}+3.[\frac{n}{3}]-n+1)}$$

$$\beta_{n+2} \;=\; a^{\frac{1}{2}.(F_{n+1}-3.[\frac{n+2}{3}]+n+1)}.b^{\frac{1}{2}.(F_{n+1}+3.[\frac{n+2}{3}]-n-1)}c^{\frac{1}{2}.(F_{n+2}+3.[\frac{n}{3}]-n+1)}$$
$$.d^{\frac{1}{2}.(F_{n+2}-3.[\frac{n}{3}]+n-1)}$$

- Scheme II:

$$\alpha_{n+2} \;=\; a^{\frac{1}{2}.(F_{n+1}+\psi(n+2))}.b^{\frac{1}{2}.(F_{n+1}\psi(n+5))}.c^{\frac{1}{2}.(F_{n+2}+\psi(n))}.d^{\frac{1}{2}.(F_{n+2}+\psi(n+3))}$$
$$\beta_n \;=\; a^{\frac{1}{2}.(F_{n+1}+\psi(n+5))}.b^{\frac{1}{2}.(F_{n+1}\psi(n+2))}.c^{\frac{1}{2}.(F_n+\psi(n+3))}.d^{\frac{1}{2}.(F_n+\psi(n))}$$

where ψ is the integer function defined in 4.1.

- Scheme III:

$$\alpha_{n+2} = a^{\sigma(n).F_{n+1}}.b^{\sigma(n+1).F_{n+1}}.c^{\sigma(n+1).F_{n+2}}.d^{\sigma(n).F_{n+2}}$$
$$\beta_{n+2} = a^{\sigma(n+1).F_{n+1}}.b^{\sigma(n).F_{n+1}}.c^{\sigma(n).F_{n+2}}.d^{\sigma(n+1).F_{n+2}},$$

where σ is an integer function defined for every $k \leq 0$ by:

m	0	1
$\sigma(2k+m)$	1	0

- Scheme IV:

$$\alpha_{n+2} = a^{F_{n+1}}.c^{F_{n+2}}$$

$$\beta_{n+2} = b^{F_{n+1}}.d^{F_{n+2}}.$$

4.2 Fibonacci sequence via arithmetic progression

The idea for this research was generated by Marchisotto's paper [23]. Thus we borrowed the first part of its title and invited colleagues to prepare a series of papers under the first part of this title.

Here we shall discuss an approach for an interpretation of the Fibonacci sequence as an arithmetic progression. The reasoning for this is the fact that there is a relation between the way of generating the Fibonacci sequence and the way of generating the arithmetic progression. On the other hand, obviously, the Fibonacci sequence is not an ordinary arithmetic progression. Thus we can construct a new type of progression which will include both the ordinary arithmetic progression, and the Fibonacci sequences (the classical one and its generalizations.

Let $f : N \to R$ be a fixed function, where N and R are the sets of the natural and real numbers, respectively, and a be a fixed real number. The sequence

$$a, a + f(1), a + f(2), ..., a + f(k), ... \tag{4.1}$$

we shall call an A-progression (from 'arithmetic progression').

Obviously, if $a_k = a + f(k)$ is its k-th member, then

$$\sum_{k=0}^{n} a_k = (n+1).a + \sum_{k=0}^{n} f(k).$$

When $f(k) = k.d$ for the fixed real number d we obtain from (4.1) the ordinary arithmetic progression.

When $a = 0$ and f is the function defined by:

$$f(1) = 1, f(2) = 1, f(k+2) = f(k+1) + f(k) \text{ for } k \geq 1,$$

we obtain from (4.1) the ordinary Fibonacci sequence. Therefore, the ordinary Fibonacci sequence can be represented by an A-progression. We shall show that some of the generalizations of this sequence can be represented by an A-progression, too. When a and b are fixed real numbers and f is a function defined by

$$f(1) = b - a, f(2) = b, f(k+2) = f(k+1) + f(k) + a,$$

we obtain from (4.1) the generalized Fibonacci sequence $a, b, a + b, a + 2.b$, $2.a + 3.b, \ldots$.

When a, b and c are fixed real numbers and f is a function defined by

$$f(1) = b - a, \quad f(2) = c - a, \quad f(3) = b + c,$$
$$f(k+3) = f(k+2) + f(k+1) + f(k) + 2.a,$$

we obtain from (4.1) the generalized Fibonacci sequence sometimes known as the Tribonacci sequence: $a, b, c, a + b + c, a + 2.b + 2.c, 2.a + 3.b + 4.c, \ldots$
When a, b, c and d are fixed real numbers, and f and g are functions defined by:

$$f(1) = -a + b, \quad f(2) = -a + c + d,$$
$$f(k+2) = g(k+1) + g(k) - a + 2.c \quad (k \geq 1)$$

$$g(1) = -c + d, \quad g(2) = a + b - c,$$
$$g(k+2) = f(k+1) + f(k) + 2.a - c \quad (k \geq 1)$$

we obtain from (4.1) the first of the generalizations of the Fibonacci sequence from Chapter 2. When for the same a, b, c and d

$$f(1) = -a + b, \quad f(2) = -a + b + c,$$

$$f(k + 2) = f(k + 1) + g(k) + c \ (k \geq 1)$$

$$g(1) = -c + d, \quad g(2) = a - c + d,$$

$$g(k + 2) = g(k + 1) + f(k) + a \ (k \geq 1)$$

we obtain from (4.1) the second of the generalizations of the Fibonacci sequence from Chapter 2.

The above idea for combination of elements of different mathematical areas with Fibonacci numbers founded a realization in the following short research, too.

Let $\{\alpha_i\}_{i=0}^{\infty}$ be a sequence with real numbers. We can construct a new sequence $\{\beta_i\}_{i=0}^{\infty}$ related to the first one, which is an analogy (and extension) of the arithmetic progression, following the scheme:

$$b_0 = a_0$$

$$b_k = b_{k-1} + \sum_{i=1}^{k} a_i. \tag{4.2}$$

For example, if $a_0 = 0, a_1 = a_2 = ... = 1$, we obtain the sequence $b_0 = 0, b_1 = 1, b_2 = 3, ..., b_k = k.(k + 1)/2 \equiv t_k$ (k-th triangle number) for $1 \leq k$.

Let $S_n = \sum_{k=1}^{n} b_k$.

The following assertion can be proved directly by induction:

Theorem 4.1: For every natural number n:

$$(a) \; b_n \; = \; a_0 + \sum_{k=1}^{n} (n+1-k).a_k,$$

$$(b) \; S_n \; = \; n.a_0 + \sum_{k=1}^{n} t_{n+1-k}.a_k. \tag{4.3}$$

We can see that $d_k = b_k - b_{k-1} = \sum_{i=1}^{k} a_i$ and $d_k - d_{k-1} = a_k$.

Therefore we obtain a situation which is analogous to acceleration in mechanics (in the sense of a velocity of a velocity). In the particular case, when $a_2 = a_3 = ... = 0$, we obtain the ordinary arithmetic progression.

The extension of the concept 'arithmetic progression' introduced in [3], which has the b-form of the following sequence:

$$b, b + d, b + 2.d, ..., b + p.d, b + p.d + e, b + p.d + 2.e, ...,$$
$$b + p.d + q.e, b + (p+1).d + q.e, ..., b + 2.p.d + q.e, b + 2.p.d + (q+1).e, ...$$

also can be represented in the above form by the a-sequence

$$\underbrace{b, d, d, ..., d}_{p \; times}, \underbrace{e, e, ..., e}_{q \; times}, \underbrace{d, d, ..., d}_{p \; times}, e, ...$$

When a sequence $\{\beta_i\}_{i=0}^{\infty}$ is given, we can construct the sequence $\{\alpha_i\}_{i=0}^{\infty}$ from the formulae (see (4.2)):

$$a_0 \; = \; b_0$$

$$a_k \; = \; b_k - \sum_{i=1}^{k} (n+1-k).a_i, \tag{4.4}$$

where a_i are previously calculated members of $\{\alpha_i\}_{i=0}^{\infty}$.

Therefore, we can define a function F, which juxtaposes to the sequence $\{\alpha_i\}_{i=0}^{\infty}$ the sequence $\{\beta_i\}_{i=0}^{\infty}$, or briefly, $F(a) = b$. If all members of the sequence b are members of sequence a, after a finite member of initial members, then we say that b is a sequence autogenerated by a. It can also be easily seen that the sequence $\{\alpha_i\}_{i=0}^{\infty}$ for which $a_i = 0$ $(0 \leq i < \infty)$ is the unique fixed point of F.

As shown above, the ordinary arithmetic progression is not autogenerated (in the general case). Below we shall construct a sequence which is autogenerate in a special sense.

Let $\alpha_0 = 1, \alpha_1 = 0, \alpha_2 = 1, \alpha_3 = 0, \alpha_4 = 1, \alpha_5 = 1, \alpha_6 = 2, \alpha_7 = 3$, etc. (after the first 3 elements, all other members of this sequence are the members of the Fibonacci sequence). Then the sequence $\{\beta_i\}_{i=0}^{\infty}$ has the form $1, 1, 2, 3, ...$, i.e. the same Fibonacci sequence without its 0-th member. We can construct another a- sequence which generates the Fibonacci sequence as its b-sequence.

Let $\alpha_0 = 0, \alpha_1 = 1, \alpha_2 = -1, \alpha_3 = 1, \alpha_4 = 0, \alpha_5 = 1, \alpha_6 = 1, \alpha_7 = 2$, etc. (after the first four elements, all other members of this sequence are the members of the Fibonacci sequence). Therefore, the Fibonacci sequence $\{F_i\}_{i=0}^{\infty}$ is an autogenerated one. From here it can easily be seen that the following equality (see [20]) is valid for every natural number n:

$$F_{n+4} = n + 3 + \sum_{k=1}^{n} (n + 1 - k).F_k.$$

Bibliography

Part A, Section 1

Coupled Recurrence Relations

1. ANDO S., Hayashi M. 1997: Counting the number of equivalence classes of (m,F) sequences and their generalizations. *The Fibonacci Quarterly* **35**, No. 1, 3–8.
2. ATANASSOV K. 1985: An arithmetic function and some of its applications. *Bulletin of Number Theory and Related Topics* Vol. IX, No. 1, 18–27.
3. ATANASSOV K. 1986: On the generalized arithmetical and geomethrical progressions. *Bulletin of Number Theory and Related Topics* Vol. X, No. 1, 8–18.
4. ATANASSOV K. 1986: On a second new generalization of the Fibonacci sequence. *The Fibonacci Quarterly* **24** No. 4, 362–365.
5. ATANASSOV K. 1989: On a generalization of the Fibonacci sequence in the case of three sequences. *The Fibonacci Quarterly* **27**, No. 1, 7–10.
6. ATANASSOV K. 1989: Remark on variants of Fibonacci squares. *Bulletin of Number Theory and Related Topics* Vol. XIII, 25–27.
7. ATANASSOV K. 1989: A remark on a Fibonacci plane. *Bulletin of Number Theory and Related Topics* Vol. XIII, 69–71.
8. ATANASSOV K. 1995: Remark on a new direction for a generalization of the Fibonacci sequence., *The Fibonacci Quarterly* **33**, No. 3, 249–250.
9. ATANASSOV K., Atanassova L., Sasselov D. 1985: A new perspective to the generalization of the Fibonacci sequence. *The Fibonacci Quarterly* **23**, No. 1, 21–28.
10. ATANASSOV K., Hlebarova J., Mihov S. 1992: Recurrent formulas of the generalized Fibonacci and Tribonacci sequences. *The Fibonacci Quarterly* **30**, No. 1, 77–79.
11. ATANASSOV K., Shannon A., Turner J. 1995: The generation

of trees from coupled third order recurrence relations. *Discrete Mathematics and Applications* (S. Shtrakov and I. Mirchev, Eds.) , Research in mathematics **5**, Blagoevgrad, 46–56.

12. BICKNELL-Johnson M., V. Hoggatt, Jr. 1972: *A primer for the Fibonacci numbers.* Santa Clara, California, The Fibonacci Association.

13. CHIARELLA, C., Shannon, A. G. 1986: An example of diabetes compartment modelling. *Mathematical Modelling* **7**, 1239–1244.

14. GEORGIEV P., Atanassov K. 1992. On one generalization of the Fibonacci sequence. Part I: Matrix representation, *Bulletin of Number Theory and Related Topics* Vol. XVI, 67–73.

15. GEORGIEV P., Atanassov K. 1995: On one generalization of the Fibonacci sequence. Part II: Some relations with arbitrary initial values. *Bulletin of Number Theory and Related Topics* Vol. XVI, 75–82.

16. GEORGIEV P., Atanassov K. 1992: On one generalization of the Fibonacci sequence. Part III: Some relations with fixed initial values. *Bulletin of Number Theory and Related Topics* Vol. XVI, 83–92.

17. GEORGIEV P., Atanassov K. 1996: On one generalization of the Fibonacci sequence. Part IV: Multiplicity roots of the characteristic equation. *Notes on Number Theory and Discrete Mathematics* **2**, No. 4, 3–7.

18. GEORGIEV P., Atanassov K. 1996: On one generalization of the Fibonacci sequence. Part V: Some examples. *Notes on Number Theory and Discrete Mathematics* **2**, No. 4, 8–13.

19. GEORGIEV P., Atanassov K. 1996: On one generalization of the Fibonacci sequence. Part VI: Some other examples. *Notes on Number Theory and Discrete Mathematics* **2**, No. 4, 14–17.

20. HOGGATT V. E. Jr. 1969: *Fibonacci and Lucas Numbers.* Boston: Houghton Mifflin Cp.

21. HOPE P. 1995: Exponential growth of random Fibonacci sequences. *The Fibonacci Quarterly* **33**, No. 2, 164–168.

22. LEE J.-Z., Lee J.-S. 1987: Some properties of the generalization of the Fibonacci sequence. *The Fibonacci Quarterly* **25**, No. 2, 111–117.

23. MARCHISOTTO E. 1993: Connections in mathematics: an introduction to Fibonacci via Pythagoras. *The Fibonacci Quarterly* **31**, No. 1, 21–27.

24. RANDIC M., Morales D.A., Araujo O. 1996: Higher-order Fibonacci numbers. *Journal of Mathematical Chemistry* **20**, No.1-2, pp.79–94.

25. SHANNON, A.G. 1983: Intersections of second order linear recursive sequences. *The Fibonacci Quarterly* **21**, Non. 1, 6–12.

26. SHANNON A., Turner J., Atanassov K. Nov. 1991: A generalized tableau associated with colored convolution trees. *Discrete Mathematics* **92**, No. 1–3, 329–340.

27. SHANNON A., Melham R. 1993: Carlitz generalizations of Lucas and Lehmer sequences. *The Fibonacci Quarterly* **31**, No. 2, 105–111.

28. SPICKERMAN W., Joyner R., Creech R. 1992; On the (2, F) generalizations of the Fibonacci sequence. *The Fibonacci Quarterly* **30**, No. 4, 310–314.

29. SPICKERMAN W., Creech R., Joyner R. 1993: On the structure of the set of difference systems defining the (3, F) generalized Fibonacci sequence. *The Fibonacci Quarterly* **31**, No. 4, 333–337.

30. SPICKERMAN W, Creech R., Joyner R. 1995: On the (3, F) generalizations of the Fibonacci sequence. *The Fibonacci Quarterly* **33**, No. 1, 9–12.

31. SPICKERMAN W., Creech R. 1997: The (2, T) generalized Fibonacci sequences. *The Fibonacci Quarterly* **35**, No. 4, 358–360.

32. STEIN, S. K. 1962: The intersection of Fibonacci sequences. *Michigan Mathematics Journal* **9**, 399–402.

33. TIRMAN A., Jablinski T. Jr. 1988: Identities derived on a Fibonacci multiplication table. *The Fibonacci Quarterly* **26**, No. 4, 328–331.

34. TURNER J.C., Shannon, A. G. 1989: On kth order coloured convolution trees and a generalized Zeckendorf integer representation theorem. *The Fibonacci Quarterly* **27**, No. 5, 439–447.

35. WELLS D. 1994: The Fibonacci and Lucas triangles modulo 2. *The Fibonacci Quarterly* **32**, No. 2, 111–123.

36. VIDOMENKO V. 1989: Combinatorics of planar δ-homogeneous tetra-angulations. *Kibernetika* **4** (in Russian), 64–69.

37. VOROB'EV N. 1978: *Fibonacci numbers.* Moskow, Nauka (in Russian).

PART A: NUMBER THEORETIC PERSPECTIVES

SECTION 2

NUMBER TREES

Krassimir Atanassov, Anthony Shannon and John Turner

Turner gave the name 'number trees' to tree graphs whose nodes or edges are weighted (or coloured) by numbers in systematic ways. Several papers on studies of number trees are cited in Chapter 1 and others are listed in the A2 Bibliography.

This Section introduces several new methods for analysing number trees. Tableaux of functions of nodal numbers are derived and generalized; Gray codes are used; various node sums are studied; and connections are made with Pascal-T triangles.

Chapter 1

Introduction – Turner's Number Trees

1.1 Introduction

In a paper by Turner and Beder [19, 1985], certain stochastic processes were defined and studied on a sequence of binary trees, in which the tree T_n had F_n leaf-nodes where F_n was the nth element of the Fibonacci number sequence (see Fig. 1 below). Turner gave the name *number trees* to tree graphs whose nodes and/or edges are weighted (or coloured) by numbers in systematic ways. An expository article on a variety of number trees is [25].

In a later paper Turner showed how to construct trees so that the nodes were weighted with integers from a general sequence $\{C_n\}$ using a sequential weighting method referred to as the 'drip-feed principle' [20, 1985a]. Subsequently it was shown how generalized Fibonacci numbers can be used to colour convolution trees so that the shades of the trees establish a generalization of Zeckendorf's theorem and its dual [21, 1985b]. There was also a construction which provided an illustration of the original Zeckendorf theorem, which established the completeness of the Fibonacci sequence and generated the Zeckendorf integer representations. In [11] all the Pythagorean triples were discovered and classified in new ways from a rational number tree.

In the sequence of Fibonacci convolution trees $\{T_n\}$ given in [20, 1985a], the sum of the weights assigned to the nodes of T_n is equal to the nth term of the convolution of $\{F_n\}$ and $\{C_n\}$. That is, if Ω means the sum of weights, we have

$$\Omega(T_n) = (F * C)_n = \sum_{i=1}^{n} F_i C_{n-i+1}. \tag{1.1}$$

For instance,
$$(F * F)_5 = F_1 F_5 + F_2 F_4 + F_3 F_3 + F_4 F_2 + F_5 F_1$$
$$= 5 + 3 + 4 + 3 + 5 = 20,$$

to which we shall refer later.

With the same tree construction, and a modified coloring rule, a graphical 'proof' was given of Zeckendorf's theorem, namely that every positive integer can be represented as the sum of distinct Fibonacci numbers, using no two consecutive Fibonacci numbers, and that such a representation is unique [8, 1964].

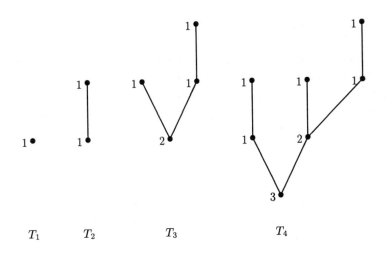

Figure 1. Number tree sequence, with Fibonacci weights

Given a sequence of colors $C = \{C_1, C_2, C_3, \ldots\}$, we construct kth order colored, rooted trees, T_n, as follows: The first k trees:

$$T_1 = C_1 \bullet; \qquad T_n = T_{n-1} \bullet\!\!-\!\!-\!\!\bullet\, C_n, \qquad n = 2, 3, \ldots, k,$$

with the root node being C_i for each of these; subsequent trees:

$$T_{n+k} = C_{n+k} \bigvee_{i=0}^{k-1} T_{n+i},$$

using the 'drip-feed' construction, in which the kth order fork operation V is to mount trees $T_n, T_{n+1}, \ldots, T_{n+k-1}$ on separate branches of a new tree with root node colored by C_{n+k}. Thus, for example, when $k = 2$, and $C = \{F_n\}$, the sequence of Fibonacci numbers, the first four second-order colored trees are as pictured in Figure 1 above.

1.2 The Tree Sequence with Weights from $F(a, b)$

Now consider the first four trees associated with $F(a, b)$, as pictured in Figure 2. The coloring sequence is the general Fibonacci one, namely, $F(a, b) = \{a, b, a + b, a + 2b, \ldots\}$.

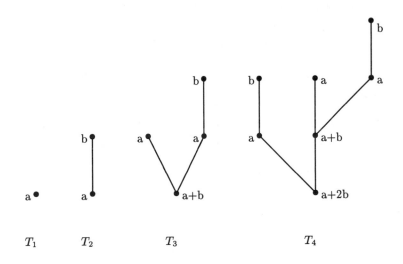

Figure 2. Number tree sequence, with general Fibonacci weights

Let (N_a, N_b) represent the number of a's and the number of b's at a given level of a tree. We may tabulate these pairs as in Table 1.

Table 1 – (N_a, N_b) at tree levels

$Level + 1 = m$	1	2	3	4	5	6
T_1	(1,0)					
T_2	(1,0)	(0,1)				
T_3	(1,1)	(2,0)	(0,1)			
T_4	(1,2)	(2,1)	(1,1)	(0,1)		
T_5	(2,3)	(2,3)	(4,1)	(2,2)	(0,1)	
T_6	(3,5)	(3,5)	(4,4)	(6,2)	(2,3)	(0,1)

If we represent the element in the nth row and mth column of this array by the vector x_{nm}, then x_{nm} satisfies the partial recurrence relation:

$$x_{nm} = x_{n-1,m-1} + x_{n-2,m-1}, \quad 1 < m < n, \quad n > 2,$$

where the addition of number pairs is elementwise, and the boundary conditions are:

$$x_{11} = x_{21} = (1,0); \quad x_{n1} = (F_{n-2}, F_{n-1}), \quad n > 2;$$
$$x_{22} = (0,1); \qquad\quad x_{nm} = (0,0), \quad m > n.$$

Chapter 2

Generalizations using Tableaux

2.1 Generalized tableaux

The tableau (of Table 1, Chapter 1) can be generalized for arbitrary k as follows: Consider x_{nm} as a k-component vector, with x_{nm} equal to the null vector when $m > n$, and

$$x_{nm} = \sum_{i=1}^{k} x_{n-i,m-1}, \quad 1 < m < n, \ n > k, \tag{2.1}$$

with

$$x_{n1} = (U_{1n}, U_{2n}, \dots, U_{kn}), \quad n = 1, 2, \dots, k,$$

where $\{U_{sn}\}, s = 1, 2, \dots, k$, are the k 'basic' sequences of order k defined by the recurrence relation

$$U_{sn} = \sum_{j=1}^{k} U_{s,n-j}, \quad n > k \tag{2.2}$$

with initial terms when $n = 1, 2, \dots, k$, $U_{sn} = \delta_{sn}$, [15], where δ_{ij} is the Kronecker delta.

When $k = 2$, we have, as before, that if we represent the element in the nth row and mth column of this array by x_{nm}, then x_{nm} satisfies the partial recurrence relation

$$x_{nm} = x_{n-1,m-1} + x_{n-2,m-1}, \quad 1 < m < n, \ n > 2,$$
$$x_{nm} = (\delta_{1m}, \delta_{2m}), \quad n = 1, 2; \ 1 \le m \le n$$

with boundary conditions $x_{n1} = (F_{n-2}, F_{n-1})$ and $x_{nm} = (0, 1)$.

As illustrations of the $\{U_{sn}\}$ we have Table 2 when $k = 3$ and Table 3 when $k = 4$.

Table 2

n	1	2	3	4	5	6	7	8	9
U_{1n}	1	0	0	1	1	2	4	7	13
U_{2n}	0	1	0	1	2	3	6	11	20
U_{3n}	0	0	1	1	2	4	7	13	24

Table 3

n	1	2	3	4	5	6	7	8	9
U_{1n}	1	0	0	0	1	1	2	4	8
U_{2n}	0	1	0	0	1	2	3	6	12
U_{3n}	0	0	1	0	1	2	4	7	14
U_{4n}	0	0	0	1	1	2	4	8	15

Various properties of $\{U_{sn}\}$ have been developed by [13]. To see more easily what follows, it is useful to continue the tree table of (N_a, N_b) for $k = 2$ (see Table 4 overleaf).

It can be observed in Tables 4 and 5 that, for $n > k$ and $m > 1$,

$$x_{nm} = \sum_{i=1}^{k} x_{n-1,m-1}. \tag{2.3}$$

As explained elsewhere [23], the rule of formation comes directly from the construction of the trees. When $k = 3$, we have the array as shown in Table 5.

The **first main result** is that for $m = 1, 2, \ldots, \lfloor \frac{n-1}{k} \rfloor$, (where $\lfloor \ \rfloor$ is the floor function):

$$x_{nm} = x_{n1}.$$

Proof and examples follow the tables overleaf.

Table 4

m	1	2	3	4	5
T_7	(5,8)	(5,8	(5,8)	(8,5)	(8,4)
T_8	(8,13)	(8,13)	(8,13)	(9,12)	(14,7)
T_9	(13,21)	(13,21)	(13,21)	(13,21)	(17,17)
T_{10}	(21,34)	(21,34)	(21,34	(21,34)	(22,33)
T_{11}	(34,55)	(34,55)	(34,55)	(34,55)	(34,55)

m	6	7	8	9	10	11
T_7	(2,4)	(0,1)				
T_8	(10,7)	(2,5)	(0,1)			
T_9	(22,11)	(12,11)	(2,6)	(0,1)		
T_{10}	(31,24)	(32,18)	(14,16)	(2,7)	(0,1)	
T_{11}	(29,50)	(53,35)	(44,29)	(16,22)	(2,8)	(0,1)

Table 5

m	1	2	3	4	5
T_1	(1,0,0)				
T_2	(1,0,0)	(0,1,0)			
T_3	(1,0,0)	(0,1,0)	(0,0,1)		
T_4	(1,1,1)	(3,0,0)	(0,2,0)	(0,0,1)	
T_5	(1,2,2)	(3,1,1)	(3,2,0)	(0,2,1)	(0,0,1)
T_6	(2,3,4)	(3,3,3)	(6,2,1)	(3,4,1)	(0,2,2)
T_7	(4,6,7)	(4,6,7)	(9,4,4)	(9,6,1)	(3,6,3)
T_8	(7,11,13)	(7,11,13)	(10,10,11)	(18,8,5)	(12,12,3)
T_9	(13,20,24)	(13,20,24)	(14,20,23)	(25,16,16)	(20,18,7)
T_{10}	(24,37,44)	(24,37,44)	(24,37,44)	(33,34,38)	(52,30,22)

m	6	7	8	9	10
T_6	(1,0,0)				
T_7	(0,2,3)	(0,0,1)			
T_8	(3,8,6)	(0,2,4)	(0,0,1)		
T_9	(15,20,8)	(3,10,10)	(0,2,5)	(0,0,1)	
T_{10}	(35,36,13)	(18,30,17)	(3,12,15)	(0,2,6)	(0,0,1)

Proof: The proof follows from induction on m by utilizing the results:

$$x_{n2} = \sum_{i=1}^{k} x_{n-i,1} = \sum_{i=1}^{k} (U_{1,n-i}, U_{2,n-i}, \ldots, U_{k,n-i})$$
$$= \left(\sum_{i=1}^{k} U_{1,n-i} \sum_{i=1}^{k} U_{2,n-i}, \ldots, \sum_{i=1}^{k} U_{k,n-1} \right)$$
$$= (U_{1,n}, U_{2,n}, \ldots, U_{k,n})$$
$$= x_{n1}, \quad \text{and so on.} \qquad \square$$

For instances (of the first main result), when k = 3,

$$x_{72} = (4, 6, 7) = (U_{17}, U_{27}, U_{37}),$$
$$x_{10,3} = (24, 37, 44) = (U_{1,10}, U_{2,10}, U_{3,10});$$

and when $k = 2$,

$$x_{52} = (2, 3) = (U_{15}, U_{25}),$$
$$x_{73} = (5, 8) = (U_{17}, U_{27}),$$
$$x_{94} = (13, 21) = (U_{19}, U_{29}),$$

in which $U_{1n} = U_{2,n-1} = F_{n-2}$ in the conventional Fibonacci notation.

The **second main result** is that for $m > \lfloor \frac{n-1}{k} \rfloor$, x_{nm} is formed from the boundary conditions

$$x_{k+1,m} = (0, 0, \ldots, k - m + 2, \ldots, 0)$$

in which the nonzero position is the $(m-1)$th; thereafter, the elements are generated by the algorithm defined by the vector difference operator Δ such that if

$$\Delta x_{nm} = x_{n+1,m+1} - x_{n,m},$$

then the sth order difference is given by

$$\Delta^s x_{n+s,n} = (0, 0, \ldots, 0, 1,), \quad \text{for } n \geq k.$$

The proof follows from the initial conditions and the ordinary recurrence relation (2.2) for $\{U_{sn}\}$ to get $x_{k+1,m}$, and then from the partial recurrence relation (2.3) for $x_{k+n,k+n-i}$. $\qquad\qquad\qquad\qquad\qquad$ □

Examples:

As examples, we have when $k = 2$,

$$x_{42} = (2,1)$$
$$\Delta x_{42} = x_{53} - x_{42} = (2,0)$$
$$x_{53} = (4,1)$$
$$\Delta^2 x_{42} = (0,1)$$
$$\Delta x_{53} = x_{64} - x_{53} = (2,1)$$
$$x_{64} = (6,2)$$
$$\Delta^2 x_{53} = (0,1)$$
$$\Delta x_{64} = x_{75} - x_{64} = (2,2)$$
$$x_{75} = (8,4)$$
$$\Delta^2 x_{64} = (0,1)$$
$$\Delta x_{75} = x_{86} - x_{75} = (2,3)$$
$$x_{86} = (10,7)$$
$$\Delta^2 x_{75} = (0,1)$$
$$\Delta x_{86} = x_{97} - x_{86} = (2,4)$$
$$x_{97} = (12,11)$$
$$\Delta^2 x_{86} = (0,1)$$
$$\Delta x_{97} = x_{10,8} - x_{97} = (2,5)$$
$$x_{10,8} = (14,16)$$
$$\Delta^2 x_{97} = (0,1)$$
$$\Delta x_{10,8} = x_{11,9} - x_{10,8} = (2,6)$$
$$x_{11,9} = (16,22)$$

When $k = 4$

$$x_{52} = (4,0,0,0)$$
$$\Delta x_{52} = (0,3,0,0)$$
$$x_{63} = (4,3,0,0)$$
$$\Delta^2 x_{52} = (0,0,2,0)$$
$$\Delta x_{63} = (0,3,2,0)$$
$$x_{74} = (4,6,2,0)$$
$$\Delta^2 x_{63} = (0,0,2,1)$$
$$\Delta x_{74} = (0,3,4,1)$$
$$x_{85} = (4,9,6,1)$$
$$\Delta^2 x_{74} = (0,0,2,2)$$
$$\Delta x_{85} = (0,3,6,3)$$
$$x_{96} = (4,12,12,4)$$
$$\Delta^2 x_{85} = (0,0,2,3)$$
$$\Delta x_{96} = (0,3,8,6)$$
$$x_{10,7} = (4,15,20,10)$$

and

$$x_{53} = (0, 3, 0, 0)$$

$$\Delta x_{53} = (0, 0, 2, 0)$$

$$x_{64} = (0, 3, 2, 0)$$

$$\Delta^2 x_{53} = (0, 0, 0, 1)$$

$$\Delta x_{64} = (0, 0, 2, 1)$$

$$x_{75} = (0, 3, 4, 1)$$

$$\Delta^2 x_{64} = (0, 0, 0, 1)$$

$$\Delta x_{75} = (0, 0, 2, 2)$$

$$x_{86} = (0, 3, 6, 3)$$

$$\Delta^2 x_{75} = (0, 0, 0, 1)$$

$$\Delta x_{86} = (0, 0, 2, 3)$$

$$x_{97} = (0, 3, 8, 6)$$

and

$$x_{54} = (0, 0, 2, 0)$$

$$\Delta x_{54} = (0, 0, 0, 1)$$

$$x_{65} = (0, 0, 2, 1)$$

$$\Delta x_{65} = (0, 0, 0, 1)$$

$$x_{76} = (0, 0, 2, 2)$$

$$\Delta x_{76} = (0, 0, 0, 1)$$

$$x_{87} = (0, 0, 2, 4).$$

Chapter 3

On Gray Codes and Coupled Recurrence Trees

3.1 Gray code of the cube, and recurrences

We use the matrices, $G, \tilde{G}$, of the Gray Code [26] of the cube and its binary complement:

$$
G = \begin{bmatrix}
0 & 0 & 0 \\
0 & 0 & 1 \\
0 & 1 & 1 \\
0 & 1 & 0 \\
1 & 1 & 0 \\
1 & 1 & 1 \\
1 & 0 & 1 \\
1 & 0 & 0
\end{bmatrix}, \quad
\tilde{G} = \begin{bmatrix}
1 & 1 & 1 \\
1 & 1 & 0 \\
1 & 0 & 0 \\
1 & 0 & 1 \\
0 & 0 & 1 \\
0 & 0 & 0 \\
0 & 1 & 0 \\
0 & 1 & 1
\end{bmatrix}
\tag{3.1}
$$

to define third-order recursive sequences $\{\alpha_n\}$ and $\{\beta_n\}$ with initial terms

$$
\begin{aligned}
\alpha_0 &= a, & \alpha_1 &= b, & \alpha_2 &= c, \\
\beta_0 &= d, & \beta_1 &= e, & \beta_2 &= f.
\end{aligned}
$$

The eight pairs of recurrence relations are defined by

$$
\begin{bmatrix} \alpha_k \\ \beta_k \end{bmatrix} = \begin{bmatrix} \bar{g}_{i\cdot} & g_{i\cdot} \\ g_{i\cdot} & \bar{g}_{i\cdot} \end{bmatrix} [\alpha_{k-3}, \alpha_{k-2}, \alpha_{k-1}, \beta_{k-3}, \beta_{k-2}, \beta_{k-1}]^T
\tag{3.2}
$$

in which $g_{i\cdot}$ is the ith row of G, $\bar{g}_{i\cdot}$ is the ith row of $\tilde{G}$, $i = 1, 2, \ldots, 8$. The Gray Codes were chosen since they permit ready generalizations for higher order recurrences. Moreover, they can be generated from a recurrence relation, and the location of any row of the matrices can be easily found with the use of a parity checker. For an example of the coupled sequences [4], applying (3.2) gives the pair (for $k \geq 3$) when $i = 2$:

$$\begin{aligned}
\alpha_k &= \beta_{k-1} + \alpha_{k-2} + \alpha_{k-3}, \\
\beta_k &= \alpha_{k-1} + \beta_{k-2} + \beta_{k-3}.
\end{aligned} \tag{3.3}$$

3.2   Third-order coloured trees

Given a sequence of colours $C = \{C_1, C_2, C_3, \ldots\}$, we construct third order coloured trees, T_n, as before:

$$T_1 = C_1 \; \bullet$$

$$T_n = T_{n-1} \; \bullet\!\!\!\rule[0.5ex]{4em}{0.4pt}\!\!\!\bullet \; C_n$$

with $C_n \bullet$ defined to be the root node in each case when $n = 2, 3, \ldots, r$, (the order of the recurrence), and

$$T_{n+3} = C_{n+3} \bigvee_{i=0}^{2} T_{n+i}, \quad n \geq 1$$

and in the 'drip-feed' construction, in which the third-order fork operation V is to mount trees T_n, T_{n+1}, T_{n+2} on separate branches of a new tree with root node at C_{n+3} for $n \geq 3$. Thus, when $n = 1$, we get

$$T_4 = \qquad \overset{\displaystyle T_1 \quad T_2 \quad T_3}{\bigvee} \atop C_4$$

We now generate graph sequences $\{T_{ijk}\}$ from the initial trees of Figure 1, and the recurrence relations (3.2).

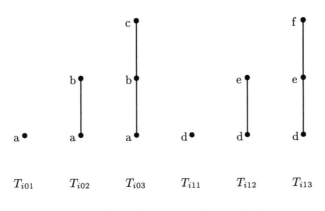

Figure 1. T_{ijk}, for $j = 0, 1; k = 1, 2, 3$.

For instance, when $i = 2$, we get the trees T_{204}, T_{205}, T_{214} and T_{215} from (3.3) as in Figure 2.

3.3 Matrix representations of coloured trees

Table 1 displays matrix representations of the trees $T_{20k}, k = 4, 5, 6, 7$. It can be observed and proved by induction on k, that N_k, the number of nodes in tree T_{ijk}, is given by the nonhomogeneous linear recurrence relation:

$$N_k = N_{k-1} + N_{k-2} + N_{k-3} + 1, \qquad k \geq 3, \tag{3.4}$$

with $N_0 = 1, N_1 = 2, N_2 = 3$. This can be solved by

$$N_k = u_k - \frac{1}{2} \tag{3.5}$$

where u_k satisfies the homogenous linear recurrence relation

$$u_k = u_{k-1} + u_{k-2} + u_{k-3} \tag{3.6}$$

with initial conditions $u_1 = \frac{3}{2}, u_2 = \frac{5}{2}, u_3 = \frac{7}{2}$.

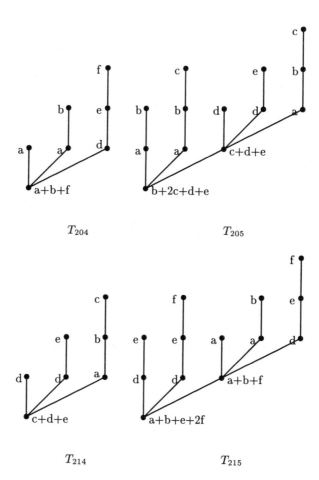

$$T_{204} \qquad\qquad T_{205}$$

$$T_{214} \qquad\qquad T_{215}$$

Figure 2. T_{2jk}, for $j = 0, 1; k = 4, 5$.

Table 1. Coupled Recurrence Trees, $T_{20k}, k = 4, 5, 6, 7$.

TREE NUMBER 204:

Number of Nodes: 7; Sum of node Weights: 15

0	0	3
0	1	1
1	1	3
5	0	0

TREE NUMBER 205:

Number of Nodes: 13; Sum of node Weights: 27

0	0	0	0	1
0	1	0	1	1
1	1	3	3	1
1	1	5	0	0
7	0	0	0	0

TREE NUMBER 206:

Number of Nodes: 24; Sum of node Weights: 68

0	0	0	0	0	0	0	0	3
0	0	0	3	3	3	3	1	1
1	0	1	1	1	1	1	1	3
1	1	1	3	3	3	5	0	0
1	5	0	0	9	0	0	0	0
15	0	0	0	0	0	0	0	0

TREE NUMBER 207:

Number of Nodes: 45; Sum of node Weights: 133

0	0	0	0	0	0	0	0	0	0	0	0	0	0	0	0	1
0	0	0	0	0	0	0	1	0	0	0	1	0	1	0	1	1
0	0	3	0	1	0	1	1	3	0	1	1	1	1	3	3	1
0	1	1	1	1	3	3	1	1	3	3	1	1	1	5	0	0
1	1	3	1	1	5	0	0	3	5	0	0	7	0	0	0	0
5	0	0	7	0	0	0	0	15	0	0	0	0	0	0	0	0
17	0	0	0	0	0	0	0	0	0	0	0	0	0	0	0	0

More generally, N_{kl}, the number of nodes at level l in T_{ijk} are displayed in Table 2 in which we see that:

$$N_{k,l} = N_{k-1,l-1} + N_{k-2,l-1} + N_{k-3,l-1} \qquad k \geq 3,\ l \geq 1. \qquad (3.7)$$

Table 2. N_{kl}, the number of nodes at level l for tree $T_{..k}$

$l =$	1	2	3	4	5	6	7	8	N_k
$k = 1$	1	0	0	0	0	0	0	0	1
2	1	1	0	0	0	0	0	0	2
3	1	1	1	0	0	0	0	0	3
4	1	3	2	1	0	0	0	0	7
5	1	3	5	3	1	0	0	0	13
6	1	3	7	8	4	1	0	0	24
7	1	3	9	14	12	5	1	0	45
8	1	3	9	21	25	17	6	1	83
9	1	3	9	25	43	41	23	7	152
10	1	3	9	27	60	80	63	30	273
11	1	3	9	27	73	128	138	92	471
12	1	3	9	27	79	176	249	224	768
13	1	3	9	27	81	212	384	450	1167
14	1	3	9	27	81	233	516	771	1641

The boundary conditions for the partial linear recurrence relation (3.7) are given by:

$$\begin{aligned}
N_{k,1} &= N_{k,k} = 1, \\
N_{k,l} &= 0,\ l > k, \\
N_{k,l} &= 3^{l-1},\ k > 3(l-1).
\end{aligned}$$

To solve (3.7), we set up the (formal) generating function

$$F_m(x) = \sum_{n=0}^{\infty} N_{n,m} x^n$$

so that

$$
\begin{aligned}
F_m(x) &= (x + x^2 + x^3)^m F_0(x) \\
&= (1 + x + x^2)^m \sum_{j=0}^{\infty} x^{m+j} \\
&= \sum_{r=0}^{2m} a_r x^r \sum_{j=0}^{\infty} x^{m+j} \\
&= \sum_{j=0}^{\infty} \sum_{r=0}^{2m} a_r x^{m+r+j} \\
&= \sum_{n=m}^{\infty} \sum_{j=m}^{n} a_{j-m} x^n
\end{aligned}
$$

in which the a_r are multinomial coefficients; on equating coefficients of x^n we get

$$N_{n,m} = \sum_{j=m}^{n} a_{j-m}.$$

Chapter 4

Studies of Node Sums on Number Trees

In Table 1 we set out $\{T_{ijk}\}$, $i = 1, 2, \ldots, 8$, $j = 0, 1$, $k = 0, 1, 2, \ldots, 12$, $a = b = c = e = 1$, $d = f = 3$. We observe that $\{T_{ijk}\}$, $i = 4, 6$, $j = 0, 1$, intersect at every fourth element, and so on. It is worth remarking that the coupled sequences $\{T_{2jk}\}$ and $\{T_{8jk}\}$ are identical, with $(\beta_k - \alpha_k)$ alternately 0 or 2. (For further discussion on the intersection of linear recurrences see ([14]; [18]). Finally, observe that the term-by-term sum over j is the same for each coupled sequence. Thus,

$$\{T_{i0k} + T_{i1k}\} = 2\{2, 1, 2, 5, 8, 15, 28, \ldots\}$$
$$= \{2u_{0,k-1} + 3u_{0,k} + 2u_{0,k+1}\}$$

where $\{u_{i,k}\}, i = 0, 1, 2$, are the three fundamental third order sequences defined by the initial terms $u_{i,k} = \delta_{i,k}$ (the Kronecker delta) for $k = 0, 1, 2$ and the recurrence relation

$$u_{i,k} = \sum_{j=1}^{3} u_{i,k-j}, \qquad k \geq 3, \ i = 0, 1, 2. \tag{4.1}$$

Note in Table 2 that

$$u_{0,k} = u_{2,k-1}$$
$$u_{1,k} = u_{0,k} + u_{2,k-2}.$$
$$T_{1,0,n} = u_{0,n} - u_{0,n-2}$$
$$T_{1,1,n} = 3u_{0,n+1} + 4u_{0,n} + u_{0,n-1}.$$

The $u_{0,n}$ can be generated from binomial coefficients [12].

Table 1. T_{ijk}, for $a = b = c = e = 1; d = f = 3; i = 1, 2, \ldots, 8; j = 0, 1$

k	1	2	3	4	5	6	7	8	9	10	11	12
T_{10k}	1	1	1	3	5	9	17	31	57	105	193	355
T_{11k}	3	1	3	7	11	21	39	71	131	241	443	815
T_{20k}	1	1	1	5	7	15	27	51	93	173	317	585
T_{21k}	3	1	3	5	9	15	29	51	95	173	319	585
T_{30k}	1	1	1	5	9	13	29	53	89	177	321	573
T_{31k}	3	1	3	5	7	17	27	49	99	169	315	597
T_{40k}	1	1	1	3	7	15	27	49	93	173	317	583
T_{41k}	3	1	3	7	9	15	29	53	95	173	319	587
T_{50k}	1	1	1	5	9	17	29	49	89	169	321	597
T_{51k}	3	1	3	5	7	13	27	53	99	177	315	573
T_{60k}	1	1	1	7	7	15	27	53	93	173	317	587
T_{61k}	3	1	3	3	9	15	29	49	95	173	319	583
T_{70k}	1	1	1	7	5	21	17	71	57	241	193	815
T_{71k}	3	1	3	3	11	9	39	31	131	105	443	355
T_{80k}	1	1	1	7	5	15	27	51	93	173	317	585
T_{81k}	3	1	3	5	9	15	29	51	95	173	319	585

Table 2. $\{U_{i,n}\}$

n	0	1	2	3	4	5	6	7	8
U_{0n}	1	0	0	1	1	2	4	7	13
U_{1n}	0	1	0	1	2	3	6	11	20
U_{2n}	0	0	1	1	2	4	7	13	24

Table 3. Sums of Node Weights

n	1	2	3	4	5	6	7
$W_{1,0,n}$	1	2	3	9	19	40	85
$W_{1,1,n}$	3	4	7	21	43	24	195
$W_{2,0,n}$	1	2	3	15	27	68	133
$W_{2,1,n}$	3	4	7	15	35	64	147

$$u_{0,n+3} = \sum_{m=0}^{\lfloor n/2 \rfloor} \sum_{r=0}^{\lfloor n/3 \rfloor} \binom{n-m-2r}{m+r} \binom{m+r}{r}. \tag{4.2}$$

Table 3 is an array of the sums of the node weights for the two sets of trees when $a = b = c = e = 1, d = f = 3$ again. It can be seen that the sum of the weights of the nodes in tree $T_{i,j,k}$ is given by $W_{i,j,k}$, where

$$W_{i,0,k} = W_{i,g_{i3},k-1} + W_{i,g_{i2},k-2} + W_{i,g_{i1},k-3} + T_{i,0,k} \tag{4.3}$$

$$W_{i,1,k} = W_{i,\bar{g}_{i3},k-1} + W_{i,\bar{g}_{i2},k-2} + W_{i,\bar{g}_{i1},k-3} + T_{i,0,k} \tag{4.4}$$

This is in fact a generalization of the result for second order sequences [16]. There the sum of the weights of the trees coloured at the nodes by the Fibonacci numbers, F_n, is given by the convolution

$$\Omega(T_n) = \sum_{i=1}^{n} F_i F_{n-i+1}.$$

Now the recurrence relation [9] for the convolution Fibonacci numbers $F_n^{(1)}$ is

$$F_n^{(1)} = F_{n-1}^{(1)} + F_{n-2}^{(1)} + F_{n-1}, \tag{4.5}$$

and the $F_n^{(1)}$ are the node weight sums $\Omega(T_n)$ above. A solution of (4.5) in terms of the Fibonacci numbers is

$$F_n^{(1)} = \frac{(n-1)F_{n+1} + (n+1)F_{n-1}}{5}.$$

Bicknell-Johnson [7] has also established that for the third order convolution numbers $G_n^{(1)}$:

$$G_n^{(1)} = G_{n-1}^{(1)} + G_{n-2}^{(1)} + G_{n-3}^{(1)} + G_{n-1}, \tag{4.6}$$

where $G_n = u_{0,n+2}$. A solution of (4.6) in terms of the tribonacci numbers $\{G_n^{(1)}\} = \{0, 1, 1, 2, 4, \ldots\}$ is

$$G_{n+1}^{(1)} = \frac{3nG_{n+1} + (7n + 12)G_n + 2(n + 1)G_{n-1}}{22}.$$

This formula is easily modified to deal with tribonacci sequences which begin with sequences with initial values other than $0, 1, 1$. It can be seen that (4.3) and (4.4) have similar forms to (4.5) and (4.6). As an illustration of (4.3) observe that

$$W_{2,1,6} + W_{2,0,5} + W_{2,0,4} + T_{2,0,7} = 64 + 27 + 15 + 27 = 133 = W_{2,0,7},$$

so that the $W_{i,j,k}$ are convolutions of the tree numbers.

Chapter 5

Connections with Pascal–T Triangles

Turner ([21], [22]) has defined the level counting function

$$L = \binom{n}{m \mid i}$$

as the number of nodes in T_n which at level m are colored C_i, where T_n is the tree coloured by integers of sequence $C = \{C_1, C_2, C_3, \ldots\}$.

One of the results proved is that

$$\binom{n}{m \mid 1} = \sum_{j=1}^{k} \binom{n-j}{m-1 \mid 1}. \tag{5.1}$$

It is also shown in effect that

$$U_{k,k+n} = \sum_{m=0}^{n} \binom{n}{m \mid 1}.$$

Thus

$$U_{k,k+n} = \sum_{m=0}^{n} \sum_{j=1}^{k} \binom{n-j}{m-1 \mid 1}. \tag{5.2}$$

It is also worth noting that (5.1) has the same form as (2.2). Now $U_{k,k+n}$ is, in the terminology of Macmahon [10], the homogeneous product sum of weight n of the zeros $\alpha_j, j = 1, 2, \ldots, k$, assumed distinct, of the auxiliary polynomial, $f(x)$, associated with the linear recurrence relation for $\{U_{k,k+n}\}$. Shannon and Horadam [16] have proved that formally

$$\sum_{n=1}^{\infty} U_{k,k+n} x^n = \sum_{m=0}^{n} \left(x^k f \left(\frac{1}{x} \right) \right)^{-1}.$$

Thus if we expand the right–hand side of

$$\sum_{n=1}^{\infty} U_{k,k+n} x^n = \frac{1}{1 - x - x^2 - \ldots - x^k}$$

by the multinomial theorem and equate corresponding coefficients of powers of x we get

$$U_{k,k+n} = \sum_{\sum_i \lambda_i = n} \frac{(\sum \lambda_i)!}{\lambda_1! \lambda_2! \ldots \lambda_k!} \tag{5.3}$$

which agrees with the analogous result in Macmahon. This is worth noting because Turner [22] has shown that the $\binom{m}{n|1}$ are multinomial coefficients generated from $x(x + x^2 + x^3 + \ldots + x^k)^m$. For example,

$$U_{2,2+n} = \sum_{\sum_i \lambda_i = n} \frac{(\sum \lambda_i)!}{\lambda_1! \lambda_2!} = \sum_{s+2m=n} \binom{n-m}{m}$$

where $\lambda_1 = s$ and $\lambda_2 = m$, as in Barakat [6], and

$$U_{3,3+n} = \sum_{\sum_i \lambda_i = n} \frac{(\sum \lambda_i)!}{\lambda_1! \lambda_2! \lambda_3!} = \sum_{s+2m+3t=n} \frac{(n-m-2t)}{s! \, m! \, t!}$$

$$= \sum_{s+2m+3t} \binom{n-m-2t}{m+t} \binom{m+t}{t}$$

where $\lambda_1 = s, \lambda_2 = m$ and $\lambda_3 = t$, as in Shannon [12].

We can also develop trees for other generalizations. For instance, Atanassov ([3], [1]) defines 2-F-sequences:

$$\alpha_{n+2} = \beta_{n+1} + \beta_n, \qquad \beta_{n+2} = \alpha_{n+1} + \alpha_n, \qquad n \geq 0 \tag{5.4}$$

with $\alpha_0 = a$, $\alpha_1 = b$, $\beta_0 = c$, $\beta_1 = d$ fixed real numbers. The trees for this scheme are shown in Figure 1.

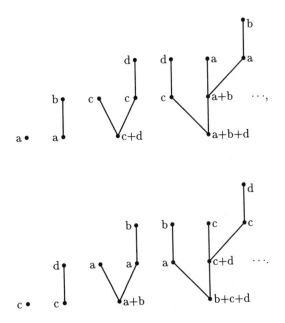

Figure 1. Tree sequences with coupled colourings

Similarly, there are 7 basic 3-F-sequences, two of which are defined by (5.5a) and (5.5b). These are studied in (Atanassov [2]).

$$\alpha_{n+2} = \gamma_{n+1} + \gamma_n,$$
$$\beta_{n+2} = \alpha_{n+1} + \alpha_n,$$
$$\gamma_{n+2} = \beta_{n+1} + \beta_n. \qquad (5.5a)$$

$$\alpha_{n+2} = \beta_{n+1} + \gamma_n,$$
$$\beta_{n+2} = \alpha_{n+1} + \alpha_n,$$
$$\gamma_{n+2} = \gamma_{n+1} + \beta_n. \qquad (5.5b)$$

These trees all have the same structure as the Fibonacci convolution trees, but their node colourings are different since their colouring rules are determined by coupled recurrences such as those of (5.4) and (5.5).

One simple illustration of how studies of the colours arising on the trees lead to interesting tableaux with Fibonacci properties is the following: For the two tree sequences S_1 and S_2 (say) from the 2-F scheme, we may compute the total weight (i.e. sum of the node colours) for each tree. For example, the fourth tree in sequence S_1 has weight $4a + 3b + 1c + 2d$. Then we may tabulate the coefficients of a, b, c, d, for each sequence (as shown up to the seventh tree in Tables 1 and 2).

Table 1. S_1 Coefficients

Tree	a	b	c	d	Σ
T_1	1	0	0	0	1
T_2	1	1	0	0	2
T_3	0	0	3	2	5
T_4	4	3	1	2	10
T_5	5	6	5	4	20
T_6	7	8	11	12	38
T_7	19	20	14	18	71

Table 2. S_2 Coefficients

Tree	a	b	c	d	Σ
T_1	0	0	1	0	1
T_2	0	0	1	1	2
T_3	3	2	0	0	5
T_4	1	2	4	3	10
T_5	5	4	5	6	20
T_6	11	12	7	8	38
T_7	14	18	19	20	71

Table 3. Total weights $\mathbf{W}_n^{(1)} + \mathbf{W}_n^{(2)}$, where $\mathbf{W}_n^{(i)}$ is the weight of $\mathbf{T}_n$ in sequence $\mathbf{S}_i$.

n	a	b	c	d	Σ_n
1	1	0	1	0	2
2	1	1	1	1	4
3	3	2	3	2	10
4	5	5	5	5	20
5	10	10	10	10	40
6	18	20	18	20	76
7	33	38	33	38	142

As we expect from the manner in which the trees were coloured (following (4.1)), the table for S_2 is table S_1 with its columns permuted thus: $(a, c)(b, d)$. Note that the sequence of row sums is

$$\{1, 2, 5, 10, \ldots, \} = \{(F * F)\},$$

the convolution of the Fibonacci sequence with itself as in (1.1).

If we add Table 1 and Table 2, elementwise, we get Table 3; and we see that the sum of the weights of the nth trees from the two sequences is:

$$W_n^{(1)} + W_n^{(2)} = U_n(a+c) + V_n(b+d), \quad \text{where}$$

$$\{U_n\} = 1,1,3,5,10,18,33,\ldots \quad and \quad \{V_n\} = 0,1,2,5,10,20,38,\ldots.$$

Now $V_n = (F * F)_{n-1}$ (proof given below); and $U_n + V_n = (F * F)_n$ (since $\sum_n$ in the table is $2(F * F)_n$); therefore $U_n = (F * F)_n - (F * F)_{n-1}$.

In Hoggatt and Bicknell-Johnson [9] the following identity for the Fibonacci convolution term is given:

$$5(F * F)_{n-1} = (n+1)F_{n-1} + (n-1)F_{n+1}.$$

Using this we obtain:

$$\begin{aligned}
V_n &= \tfrac{1}{5}[(n+1)F_{n-1} + (n-1)F_{n+1}]; \text{ and so} \\
5U_n &= [(n+2)F_n + nF_{n+2}] - [(n+1)F_{n-1} + (n-1)F_{n+1}] \\
&= (n+1)(F_n + F_{n-2}) + F_{n+1}.
\end{aligned}$$

Therefore

$$U_n = \frac{1}{5}[(n+1)L_{n-1} + F_{n+1}], \quad \text{where } L_{n-1} \text{ is a Lucas number.}$$

We finally prove the convolution forms given above for U_n, V_n thus:

Proof: It was established in (Turner [20]) that if a single sequence of the convolution trees is coloured sequentially, using colour C_n of a sequence $\{C_n\}$ to colour the root node of T_n, and mounting the previously coloured T_{n-1} and T_{n-2} on the fork, then the weight of T_n is $(F * C)_n$.

Now the general term of $\sum_n$ (in table 3) is obtained by setting $a = b = c = d = 1$; in that event, both S_1 and S_2 are Fibonacci convolution trees (i.e., $C = F$ in both cases), so $\sum_n = 2(F * C)_n$.

Similarly, if we set $a = 0 = c$ and $b = 1 = d$, we find that S_1 and S_2 are identical but with colour sequences $\{F_{n-1}\}$; and then $U_n \cdot 0 + V_n \cdot 2 = 2(F * F)_{n-1}$, giving the required form of V_n. $\square$

Bibliography

Part A, Section 2

Number Trees

1. ATANASSOV, K. T. 1986: On a second new generalization of the Fibonacci sequence. *The Fibonacci Quarterly.* **24(4)**, 362–365.
2. ATANASSOV, K. T. 1989: On a generalization of the Fibonacci sequence in the case of three sequences. *The Fibonacci Quarterly.* **27(1)**, 7–10.
3. ATANASSOV, K. T., Atanassova, L. C. and Saselov, D. D. 1985: A new perspective to the generalization of the Fibonacci sequence. *The Fibonacci Quarterly.* **23(1)** pp 21–28.
4. ATANASSOV, K. T., Hlebarska, J. and Mihov, S. 1992: Recurrent formulas of the generalized Fibonacci and triboncci sequences. *The Fibonacci Quarterly.* **30(4)**, 77–79.
5. ATANASSOVA, K. T., Shannon, A. G. and Turner, J. C. 1995: The Generation of trees from coupled third order recurrence relations. in Shtrakov, S. and Mirchev. Iv. (eds.) *Discrete Mathematics and Applications* (Research in Mathematics Volume 5), Neofit Rilski University, Blagoevgrad, 46–56.
6. BARAKAT, R. 1964: The matrix operation e^x and the Lucas polynomials. *Journal of Mathematics and Physics.* **43(4)**, 332–335.
7. BICKNELL-Johnson, M. June 22, 1988: Private communication to J.C. Turner.
8. BROWN Jr, J. L. 1964: Zeckendorf's Theorem and some applications. *The Fibonacci Quarterly.* **2(3)**, 163–168.
9. HOGGATT Jr, V. E. and Bicknell-Johnson, M. 1977: Fibonacci convolution sequences. *The Fibonacci Quarterly.* **15(2)**, 117–122.
10. MACMAHON, P. A. 1915: *Combinatory Analysis.* Vol I. Cambridge University Press, Cambridge, 2–4.
11. SCHAAKE A.G., Turner J. C. 1989: *A New Chapter for*

Pythagorean Triples. pub. Department of Mathematics, University of Waikato, Hamilton, New Zealand.

12. SHANNON, A. G. 1972: Iterative formulas associated with third order recurrence relations. *SIAM Journal of Applied Mathematics.* **23(3)**, 364–368.

13. SHANNON, A. G. 1974: Some properties of a fundamental recursive sequence of arbitrary order. *The Fibonacci Quarterly.* **12(4)**, 327–335.

14. SHANNON, A. G. 1983: Intersections of second order linear recurrences. *The Fibonacci Quarterly.* **21(1)**, 6–12.

15. SHANNON, A. G. and Bernstein, L., 1973: The Jacobi–Perron algorithm and the algebra of recursive sequences. *Bull. Austral. Math. Soc.* **8(2)**, 261–277.

16. SHANNON, A. G. and Horadam, A. F. 1991: Generalized staggered sums. *The Fibonacci Quarterly.* **29(1)**, 47–51.

17. SHANNON, A. G., Turner, J. C. and Atanassov, K. T. 1991: A generalized tableau associated with colored convolution trees. *Discrete Mathematics,* **92.4**, 323–340.

18. STEIN, S. K. 1962: The intersection of Fibonacci sequences. *Michigan Mathematical Journal* **9(4)**, 399–402.

19. TURNER, J. C. 1985: Stochastic processes defined on tree sequences. *University of Waikato Mathematical Research Report,* No. 142.

20. TURNER, J. C. May, 1985a: Fibonacci convolution trees and integer representations. *New Zealand Mathematical Society Newletter.,* 16–22.

21. TURNER, J. C. 1985b: Basic primary number and Pascal–T triangles. *University of Waikato Mathematical Research Report,* No. 145.

22. TURNER, J. C. 1988: Fibonacci word patterns and binary sequences. *The Fibonacci Quarterly.* **26(3)**, 233–246.

23. TURNER, J. C. and Shannon, A. G. 1979: On kth order colored convolution trees and a generalized Zeckendorf representation theorem. *The Fibonacci Quarterly.* **15(2)**, 117–122.

24. TURNER, J. C. 1988: Convolution trees and Pascal–T triangles. *The Fibonacci Quarterly.* **26(3)**, 354–365.

25. TURNER, J. C. 1990: Three number trees – their growth rules and related number properties. G. E. Bergum et als. (eds.), *Applications of Fibonacci numbers.* Kluwer Ac. Pub., 335–350.

26. WILF, H. S. 1989: *Combinatorial Algorithms: An Update.* S.I.A.M., Philadelphia.

PART B: GEOMETRIC PERSPECTIVES

SECTION 1

FIBONACCI VECTOR GEOMETRY

John Turner

Fibonacci Vector Geometry is the study of properties of vectors whose coordinates are drawn from integer sequences which are generated by linear recurrence equations. Normally the vectors consist of integer triples, taken in order from an integer recurrence sequence.

The vectors can be studied geometrically, as points in $\mathbf{Z}^3$. Then sequences of vectors can be regarded as polygons (by joining up the points with line segments); such polygons lie in planes, and a variety of related geometric objects can be defined and studied.

This approach is helpful in suggesting many ways to study classes of integer vector sequences, enabling geometric theorems about them to shed light on their individual and collective properties. Conversely, geometric objects can be generated sequentially, using Fibonacci-type recurrences, and study of these objects sheds new light on the integer sequences associated with their vertices.

The author began developing these ideas and themes in 1994.

Chapter 1

Introduction and Elementary Results

1.1 Introduction

In this and the following chapters, ideas about integer vectors and vector recurrence equations are introduced. Then various integer vector sequences are studied, both algebraically and geometrically. In almost all of the work, the vectors are 3-dimensional and with integer co-ordinates. Thus the work is essentially algebraic geometry in $\mathbf{Z}^3$; and it is strongly connected with integer sequences. Sometimes the results will hold also in $\mathbf{R}^3$; it will be clear from context when this is intended.

Many of the examples and theorems will relate to the doubly infinite sequence of the Fibonacci numbers $..., F_{-n}, ..., -1, 1, 0, 1, 1, 2, 3, 5, ..., F_n, ...$; or to the Lucas numbers which are defined by $L_n \equiv F_{n-1} + F_{n+1}$; or to generalised Fibonacci numbers which are defined next.

> **Definition 1.1:**
> A general Fibonacci sequence of integers is defined by $\{G_n\}$, with $G_1 = a$, $G_2 = b$, and a, b and $n \in \mathbf{Z}$, and with terms of this sequence satisfying the linear recurrence:
> $$G_{n+2} = G_{n+1} + G_n.$$
> Thus $\{G_n\} \equiv ..., a, b, a+b, 1a+2b, 2a+3b, ..., F_{n-2}a + F_{n-1}b, ...$ where a and b are integers. The sequence $\{G_n\}$ will sometimes be referred to simply by $F(a, b)$.

Much of the vector geometry to be described deals with the so-called *Fibonacci vectors*, which we define as follows:

Definition 1.2:
The nth *Fibonacci vector* is $\mathbf{F}_n \equiv (F_{n-1}, F_n, F_{n+1})$.
The nth *Lucas vector* is $\mathbf{L}_n \equiv (L_{n-1}, L_n, L_{n+1})$.
The nth *general Fibonacci vector* is $\mathbf{G}_n \equiv (G_{n-1}, G_n, G_{n+1})$.

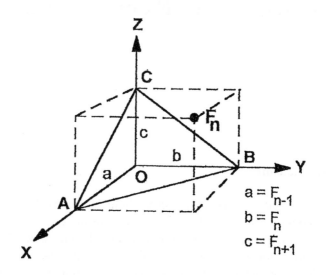

Figure 1. The Fibonacci vector point

The work of this Section evolved from ideas advanced by the author (Turner) in several papers which developed mathematical theories about sequences of integer pairs or triples by linking them with geometric diagrams or graphs. For example, in [27] rational numbers were studied as integer pairs, arranging them in a special way on the nodes of a binary tree, producing what was called 'enteger geometry'. The chapters in this book deal with integer triples obtained from well-known integer sequences and treated as vectors as defined in (1.2) above. The ideas presented combine in a variety of ways work on integer sequences (known as Fibonacci mathematics) with well-known ideas of three-dimensional geometry: so this new subject-matter would seem to deserve the title *Fibonacci Vector Geometry*.

It is particularly helpful in suggesting ways to study classes of integer sequences by depicting them as geometric objects, and enabling geometric theorems to shed light on their individual and collective properties. Conversely, geometric objects can be generated sequentially, using Fibonacci-type recurrences, and study of these objects sheds new light on the integer sequences associated with their vertices.

The author (Turner) first presented some of these ideas at a Conference of the Australian Mathematical Society, held in Hobart, Tasmania, in January 1995. A further introductory paper was presented to the Seventh International Conference on Fibonacci Numbers and Their Applications, which took place in Graz, Austria, in July, 1996. This paper [30] subsequently appeared in the Graz Conference Proceedings. The chapters in this book combine and extend material introduced in those two talks.

Some similar geometric work has been done by others, but mostly in two dimensions in the XY-plane, with point-coordinates being consecutive pairs of elements from Fibonacci sequences [4],[15],[20],[10].

Work of a somewhat different nature, on geometric tessellations involving the golden section, arose as an off-shoot from the Fibonacci vector geometry. Some of this so-called *goldpoint geometry* is presented by the author and a co-developer, Vassia Anatassov, in later chapters.

Most previous work on Fibonacci tessellations of spaces has focussed on tiling [2],[3],[9],[11],[13], as has some previous work linking Fibonacci numbers with geometry [5]. Regular tiling, such as that investigated below, has generally concentrated on polyominoes [17],[21],[24],[34].

At first thought, taking the idea of Fibonacci vectors into three dimensions would seem somewhat pointless in view of the linear dependency between their coordinates; but the author quickly found it to have many appealing consequences. For example, each member of a large class of linear second-order recurrence equations determines (or corresponds to) a plane in $\mathbf{R}^3$: each integer sequence generated by a member equation (with given starting values) determines a vector polygon which lies in the corresponding plane; indeed, the set of such polygons partitions the integer points of the plane. We shall see in chapter 3 that the polygons determined by the basic Fibonacci vector sequences lie in a plane which has a honeycomb of integer points; and much interesting geometry concerning them can be discovered. Whereas, for example, the polygons and geometry of Pell vector sequences occur in a different plane. Comparisons and links between these Fibonacci and Pell geometries make interesting and fruitful studies.

1.2   The Fibonacci Vector—some Elementary Results

Before introducing linear vector recurrences, a few elementary results which derive directly from the *Fibonacci vector* of Definition 1.2 will be presented.

RESULTS I: Solutions of 'squares' equations

Figure 1 above shows the 'point' of vector $\mathbf{F}_n \equiv (F_{n-1}, F_n, F_{n+1})$ referred to Cartesian axes with origin $Q(0,0,0)$.

Surprisingly, the first geometric object suggested by this simple diagram, namely the triangle ABC, proved worthy of study. The formula for its area, given below, led immediately to interesting results about an infinite sequence of Diophantine equations.

$$\Delta_n \equiv \text{Area } \Delta ABC = \frac{1}{2}\sqrt{(F_n F_{n-1})^2 + (F_{n+1} F_{n-1})^2 + (F_n F_{n+1})^2} \quad (1.1)$$

The given formula may be checked using any text on 3D-geometry (see, e.g. [8]), or by elementary means directly from the diagram. Its easiest derivation is obtained by applying the vector product formula
$$\text{Area} = \tfrac{1}{2}|\underline{AC} \times \underline{BC}|.$$

On calculating the first six values of Δ_n, the following sequence of triangle areas was obtained: $\Delta_n = \tfrac{1}{2}\{1,\ 3,\ 7,\ 19,\ 129,\ 337,\ ...\ \}$.

It is easy to show that the ratio of consecutive terms of this sequence, Δ_{n+1}/Δ_n, tends to α^2 as n tends to infinity (where α is the Golden Mean).

It was a pleasant surprise to find that the expression under the root sign, in the formula for Δ_n, was always a perfect square. With a little algebraic manipulation, using Fibonacci identities, it can be shown that:

$$(F_n F_{n-1})^2 + (F_{n+1} F_{n-1})^2 + (F_n F_{n+1})^2$$
$$= (F_{n-1}^2 + F_n{}^2 + F_{n-1} F_n)^2 \quad (1.2)$$

Consequently, from (1.1), we can write

$$\Delta_n = \tfrac{1}{2}(F_{n-1}^2 + F_n{}^2 + F_{n-1} F_n) = \tfrac{1}{2}(F_{2n-1}^2 + F_{n-1} F_n) \quad (1.3)$$

Moreover, it is evident from (1.2) that a general solution for the 'four-squares equation' $x^2 + y^2 + z^2 = w^2$, in terms of the Fibonacci numbers, has been discovered. (N.B. It is easy to show [32] that we can replace F by L in the above equation (2), and thus obtain a similar solution to the four-squares equation, in terms of the Lucas numbers.)

The author was encouraged to look for solutions to the general m-squares equation, in terms of the Fibonacci numbers, namely for the equation:

$$x_1^2 + x_2^2 + x_3^2 + \ldots + x_{m-1}^2 = x_m^2 \tag{1.4}$$

The general solutions he found, for $m = 3, 4, 5, \ldots$, are given below. The presentation has a poetic style to it; so I was tempted to entitle it *Square Dance in Fibonacci Numbers*, an ode with an infinity of verses.

A SQUARE DANCE IN FIBONACCI NUMBERS

(1) **The equation:** $x^2 + y^2 = z^2$.

Solution:
$$\begin{aligned}
x &= F_{2n-1} + F_{n-1}F_n \\
y &= 2F_{n-1}F_n^2 F_{n+1} \\
z &= 2F_{n-1}F_n^2 F_{n+1} + 1.
\end{aligned}$$

(2) **The equation:** $x^2 + y^2 + z^2 = w^2$.

Solution:
$$\begin{aligned}
x &= F_{n-1}F_n \\
y &= F_{n-1}F_{n+1} \\
z &= F_n F_{n+1} \\
w &= F_{n-1}^2 + F_n^2 + F_{n-1}F_n.
\end{aligned}$$

(3) **The equation:** $x^2 + y^2 + z^2 + u^2 = v^2$.

Solution:
$$\begin{aligned}
x &= F_{n-1}F_n \\
y &= F_{n-1}F_{n+1} \\
z &= F_n F_{n+1} \\
u &= \tfrac{1}{2}[(F_{n-1}^2 + F_n^2 + F_{n-1}F_n)^2 - 1] \\
v &= \tfrac{1}{2}[(F_{n-1}^2 + F_n^2 + F_{n-1}F_n)^2 + 1].
\end{aligned}$$

$$\ldots \quad \ldots \quad \ldots$$
$$\ldots \quad \ldots \quad \ldots$$

[N.B. The general case follows on the next page.]

Extension to the mth case (and hence to infinity): we can use the identity $v^2 \equiv (v^2+1)^2/4-(v^2-1)^2/4$ to extend verse (3) to verse (4), and so on ad infinitum. [It is necessary to remark that the last variable (on the right-hand side of the m-squares equation) is always odd.]

RESULTS II: Geometric properties of Fibonacci triangles

Some elementary geometric results concerning the geometry of the triangle ABC, shown in Figure 1 above, now follow. They are easily confirmed, using formulae from [8], say.

(i) The Fibonacci and Lucas F-triangles ABC lie, respectively, in the following planes:

$$\frac{x}{F_{n-1}} + \frac{y}{F_n} + \frac{z}{F_{n+1}} = 1 \quad \text{and} \quad \frac{x}{L_{n-1}} + \frac{y}{L_n} + \frac{z}{L_{n+1}} = 1 .$$

(ii) Let the **normal** from the point F_n to the plane of ABC be denoted by PN. Then its length is $|PN| = 2F_{n-1}F_nF_{n+1}/(F_{n-1}^2 + F_n^2 + F_nF_{n-1})$.

(iii) The angle between the nth Fibonacci and nth Lucas F-triangles is given by:

$$\cos\theta = 4F_{2n}/\sqrt{D_F D_L} , \quad \text{where}$$

$$D_F = F_{n-1}^2 + F_n^2 + F_{n+1}^2 \text{ and } D_L = L_{n-1}^2 + L_n^2 + L_{n+1}^2 .$$

(iv) Consider two consecutive Fibonacci F-triangles. Let their respective areas be Δ_n and Δ_{n+1}. then:
  (iv)(a) **Ratio**

$$\frac{\Delta_{n+1}}{\Delta_n} = \frac{F_nF_{n+1} + F_{2n+1}}{F_{n-1}F_n + F_{2n-1}} .$$

This ratio tends to α^2 as n tends to infinity, where α is the golden ratio.
  (iv)(b) **Difference**

$$\Delta_{n+1} - \Delta_n = F_nF_{n+1}F_{2n+1} .$$

(v) Consider the sides of the general F-triangle. Each of them is the hypotenuse of a right triangle, made with two of the reference axes. Let $AB = u, BC = v, CA = w$.

(v)(a) Using Pythagoras' theorem and a Fibonacci identity we obtain:

$$u = AB = \sqrt{F_{2n-1}}, \quad v = BC = \sqrt{F_{2n+1}},$$

$$w = CA = \sqrt{F_{n-1}^2 + F_{n+1}^2} = \sqrt{L_n^2 - 2F_{n-1}F_{n+1}}.$$

It follows that, using the formula (1.1) for Δ_n which was given above, and also its standard trigonometric formula:

$$\frac{1}{2}(F_{n-1}F_n + F_{2n-1}^2) = \sqrt{s(s - \sqrt{F_{2n-1}})(s - \sqrt{F_{2n+1}})(s - \sqrt{F_{n-1}^2 + F_{n+1}^2})},$$

where s is half the sum of the sides, that is $s = (u + v + w)/2$.

(v)(b) Let θ be the angle between AB and BC.
Using the cosine formula $w^2 = v^2 + u^2 - 2vu\cos\theta$, and also direction ratios for AB and BC, we can find expressions for $\cos\theta$ in two ways. Equating these gives the following identity:

$$L_n^2 - L_{2n} = 2(F_{n-1}F_{n+1} - F_n^2).$$

(vi) If we take the origin $Q(0,0,0)$ as a fourth point joined to the vertices of the F-triangle, we have defined a tetrahedron T_n (indeed a sequence of tetrahedrons, if n is allowed to vary). The volumes are given by:

$$V_n = \text{Vol}(T_n) = \frac{1}{6} \begin{Vmatrix} 0 & 0 & 0 & 1 \\ F_{n-1} & 0 & 0 & 1 \\ 0 & F_n & 0 & 1 \\ 0 & 0 & F_{n+1} & 1 \end{Vmatrix} = \frac{1}{6}F_{n-1}F_nF_{n+1}.$$

It follows easily that:

(vi)(a) The **ratio** V_{n+1}/V_n tends to α^3 as n tends to infinity.

(vi)(b) The **difference** $V_{n+1} - V_n = \frac{1}{3}F_{n+1}F_n^2$.

Formulae similar to those given in (i) to (vi) can be given for Lucas F-triangles and tetrahedra, in terms of the Lucas numbers.

RESULTS III: Some Fibonacci Vector Identities

The third set of results is a collection of elementary identities involving Fibonacci vectors. In order to discover them it was only necessary to take well-known Fibonacci identities, and, wherever terms such as F_n occurred in the left-hand sides, replace them by $\mathbf{F}_n$; then it was usually a simple matter to determine the form of the new right-hand side, in terms of Fibonacci vectors*. We believe that some have more mathematical charm than their original ones in terms of the ordinary Fibonacci numbers. Moreover, the scope for deriving vector identities is widened by the fact that both dot-products and vector-products can enter into them too; some Fibonacci identities give rise to more than one vector identity.

To save space, we shall list the Fibonacci vector identities directly, without giving the ordinary Fibonacci identites from whence they came (these should be obvious to the reader).

$$\text{(i)} \qquad \mathbf{F}_n = \mathbf{F}_{n-1} + \mathbf{F}_{n-2}\,.$$

$$\text{(ii)} \qquad \mathbf{L}_n = \mathbf{L}_{n-1} + \mathbf{L}_{n-2}\,.$$

$$\text{(iii)} \qquad \mathbf{L}_n = \mathbf{F}_{n-1} + \mathbf{F}_{n+1}\,.$$

$$\text{(iv)} \qquad \mathbf{F}_1 + \mathbf{F}_2 + \mathbf{F}_3 + \ldots + \mathbf{F}_n = \mathbf{F}_{n+2} - \mathbf{F}_2\,.$$

$$\text{(v)} \qquad \mathbf{L}_1 + \mathbf{L}_2 + \mathbf{L}_3 + \ldots + \mathbf{L}_n = \mathbf{L}_{n+2} - \mathbf{L}_2\,.$$

The next three identities constitute another ode in three verses: this one we call *Partial Sums and Parities*.

$$\text{(vi)} \qquad \mathbf{F}_1 + \mathbf{F}_3 + \mathbf{F}_5 + \ldots + \mathbf{F}_{2n-1} = \begin{cases} L_n\mathbf{F}_n, & \text{if } n \text{ is odd;} \\ F_n\mathbf{L}_n, & \text{if } n \text{ is even.} \end{cases}$$

$$\text{(vii)} \qquad \mathbf{F}_2 + \mathbf{F}_4 + \mathbf{F}_6 + \ldots + \mathbf{F}_{2n} = \begin{cases} L_n\mathbf{F}_{n+1}, & \text{if } n \text{ is odd;} \\ F_n\mathbf{L}_{n+1}, & \text{if } n \text{ is even.} \end{cases}$$

$$\text{(viii)} \qquad \mathbf{F}_1 - \mathbf{F}_2 + \ldots + \mathbf{F}_{2n-1} - \mathbf{F}_{2n} = \begin{cases} -L_n\mathbf{F}_{n-1}, & \text{if } n \text{ is odd;} \\ -F_n\mathbf{L}_{n-1}, & \text{if } n \text{ is even.} \end{cases}$$

*N.B. Sometimes no such form emerges, and hence no corresponding identity can be given.

Some Fibonacci vector identities with multiples of 2 occurring as subscripts now follow:

(ix) $\qquad \mathbf{F}_{2n} = F_{n-1}\mathbf{F}_n + F_n\mathbf{F}_{n+1}$.

(x) $\qquad \mathbf{L}_{2n} = L_{n-1}\mathbf{F}_n + L_n\mathbf{F}_{n+1}$.

(xi) $\qquad \mathbf{F}_{2^i n} = F_{(2^i-1)n-1}\mathbf{F}_n + F_{(2^i-1)n}\mathbf{F}_{n+1}$.

When binary products of Fibonacci numbers occur in an identity, they can be replaced by scalar products, or cross products, of Fibonacci vectors. Thus sometimes more than one vector identity may be found, corresponding to a single Fibonacci number identity. (xii)–(xv) below give simple examples of products of Fibonacci vectors.

(xii) $\qquad \mathbf{F}_n.\mathbf{F}_n = F_{n+1}F_{n+2} - F_{n-1}F_{n-2}$.

(xiii) $\qquad \mathbf{F}_n.\mathbf{F}_{n+1} = F_{n-1}F_n + F_{n+1}L_{n+1}$.

(xiv) $\qquad \mathbf{F}_n.\mathbf{F}_{n+2} = F_{n+1}F_{n+3} + F_{n+4}$.

(xv) $\qquad \mathbf{F}_n \wedge \mathbf{F}_{n+1} = (-1)^n(1, 1, -1)$.

In the next chapter we define general linear recurrences for vectors, and discuss the particular case for generating Fibonacci vectors sequentially. In later chapters, we shall study further the geometry of Fibonacci vectors, finding where the vectors (regarded as points) are situated, and how they configure together in 3-space.

Chapter 2

Vector Sequences from Linear Recurrences

2.1 Integer-vector Recurrence Equations

Any recurrence equation for numbers can be turned into a recurrence equation for vectors, by the simple technique of changing number-terms into corresponding vector-terms. And then, providing the initial vectors are integer-vectors (i.e. vectors with integer coordinates), and the coefficients in the equation are suitably chosen integers, the recurrence will produce a sequence of integer-vectors. Of course, the arithmetic operations involved must also be changed into corresponding vector operations—thus the numerical + operation will become the vector + operation, and so on.

The most general vector recurrence of this kind which we wish to study is given by the following definition.

> **Definition 2.1:** A general *linear integer-vector recurrence equation* is: $\mathbf{x}_{n+2} = c\mathbf{x}_{n+1} + d\mathbf{x}_n$ (2.1)
> where c and d are integers, and $\mathbf{x}_n$ and $\mathbf{x}_{n+1}$ are integer-vectors (of the same dimension).

If the initial integer-vectors $\mathbf{x}_1$ and $\mathbf{x}_2$ are given, the recurrence (2.1) will generate an *integer-vector sequence*. This may be singly or doubly infinite.

> **Definition 2.2:** By putting $c = 1 = d$ in (2.1), we obtain the simplest recurrence, which we shall call the *Fibonacci vector recurrence*: $\mathbf{x}_{n+2} = \mathbf{x}_{n+1} + \mathbf{x}_n$ (2.2)

We next describe the general Fibonacci vector sequence, which is generated by means of the recurrence (2.2), with the initial vectors designated by **a** and **b**.

2.2 Fibonacci Vector Sequences

If we let $x_1 = \mathbf{a}$ and $x_2 = \mathbf{b}$ in (2.2), then the doubly infinite Fibonacci vector sequence generated has terms thus:

$$\dots, \mathbf{a}, \mathbf{b}, \mathbf{a} + \mathbf{b}, \mathbf{a} + 2\mathbf{b}, 2\mathbf{a} + 3\mathbf{b}, \dots, F_{n-2}\mathbf{a} + F_{n-1}\mathbf{b}, \dots \qquad (2.3)$$

It has exactly the same form as F(a,b) (see Def. 1.2), but with a,b changed to $\mathbf{a}, \mathbf{b}$ respectively. We shall use $\mathbf{G} \equiv \{\mathbf{G}_n\}$ to denote this sequence, and call it the *general Fibonacci vector sequence*.

In this book we shall require $\mathbf{a}$ and $\mathbf{b}$ to represent fixed 3-dimensional vectors, with initial points at the origin $Q(0,0,0)$ and with end-points at $A(a_1, a_2, a_3)$ and $B(b_1, b_2, b_3)$ respectively. Usually, we shall require all the coordinates to be integers. In other words, generally $\mathbf{a}$ and $\mathbf{b}$ are position vectors in $\mathbf{Z}^3$.

Example

If $\mathbf{a} = (1,2,3) = A(\text{or point } P_1)$, and $\mathbf{b} = (2,3,4) = B(\text{or point } P_2)$, then $x_3 = (3,5,7) = P_3$, $x_4 = (5,8,11) = P_4$ and etc.

We can take the sequence backwards, operating the recurrence 'to the left', obtaining terms $x_0 = (1,1,1)$, $x_{-1} = (0,1,2)$ and so on for terms $x_{-2}, x_{-3}, \dots$

We shall now study some geometric properties of sequences of the form (2.3), namely the general Fibonacci vector sequences.

2.3 Geometric Properties of G

The diagram below (see Fig. 1) shows how the sequence of vectors from (2.3) appear in space.

We note at the outset that since each vector in the sequence is of the form $m\mathbf{a} + n\mathbf{b}$, they each lie in the plane determined by the origin $Q(0,0,0)$ and the two points A and B (assuming that $\mathbf{a}$ and $\mathbf{b}$ are not collinear). We shall call this plane $\pi(\mathbf{a}, \mathbf{b})$.

We also note from (2.3) that the coefficients m and n are Fibonacci numbers.

Observations:

Several simple, but attractive, geometric observations can be inferred more or less directly from Fig. 1, and from the law of vector addition. Thus:

(i) We have assumed that **a** and **b** are not collinear. If they were, then the whole sequence would lie in the direction of **a**, if $\theta_1 = 0°$.

Whereas if $\theta_1 = 180°$, the vectors $\mathbf{G}_n$ will alternate in direction for a while, before settling into the direction of **a** or of **b**, depending on the respective values of $|\mathbf{a}|$ and $|\mathbf{b}|$. We know that $\mathbf{a} = c\mathbf{b}$, with c a negative constant: then if $|c| > \alpha = 1.6180...$ the sequence will eventually move in the direction of **a**, and otherwise it will move in the direction of **b**.

(ii) After some value of n, say $n > N$, the lengths $|\mathbf{G}_n|$ are increasing with n. N is the integer for which $|\mathbf{G}| = |F_{n-2}\mathbf{a}+F_{n-1}\mathbf{b}|$ is minimal. In general, this value is unique. The only exceptions are the basic Fibonacci vector sequence and its multiples. These each have three terms which are of equal smallest length.

(iii) The angle θ_n between $\mathbf{G}_n$ and $\mathbf{G}_{n+1}$
   (a) alternates in sense, clockwise, anti-clockwise, etc., and
   (b) decreases in size, since it is always less than the angle of the parallelogram formed on $\mathbf{G}_{n-1}$ and $\mathbf{G}_n$.

(iv) In view of (ii) and (iii) we can deduce that the vectors $\mathbf{G}_n$ tend, as $n \to \infty$, towards an upward limit ray, emanating from $Q(0,0,0)$, which we call **L**. In Section 2.4 below, in Theorem 2.1, we shall derive equations for this limit ray.

Comments:

(1) The changes in $\mathbf{G}_n$, and its convergence to **L**, as n steps through the sequence 1, 2, 3, ... are geometric analogues of the changes that occur in the ratio F_{n+1}/F_n from the Fibonacci number sequence: the values of this ratio alternate in value, being alternately above and below the golden mean α; and they tend to α with n.

(2) It is not evident from Fig. 1 that there are two other limit rays to the vectors of sequence (2.3). In fact, when $n < N$, and $n \to -\infty$, the vectors $\mathbf{G}_n$ still oscillate, but they point respectively 'left' and 'right' as

they grow in length. This happens because after $n = -1$ the coefficients F_{n-2}, F_{n-1} of $\mathbf{a}, \mathbf{b}$ respectively begin to alternate in sign. There is thus also a left-limit ray (call it $\mathbf{L'}$) and a right-limit ray (call it $\mathbf{L''}$) for the vector sequence $\mathbf{G}$. We study these two rays in section 2.4 below.

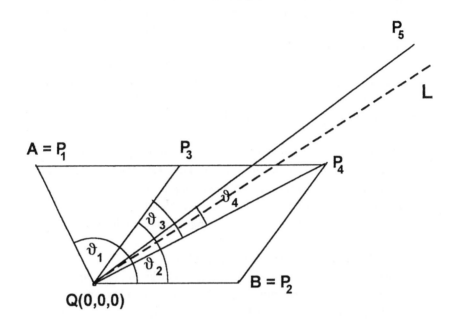

Figure 1. The Fibonacci Vector Sequence in plane $\pi(\mathbf{a}, \mathbf{b})$

We now continue with further geometric observations from Figure 1.

(v) $AP_3 = P_3P_4$, with A, P_3, P_4 being collinear points.

   Proof: QB and AP_3 form a parallelogram (by construction of the point P_3); and QB and P_3P_4 also form a parallelogram (by construction of point P_4). Hence both AP_3 and P_3P_4 are equal in length and parallel to QB.   □

(vi) It follows from (v) that parallelograms QAP_3B and $QP_2P_4P_3$ have equal areas.

Corollary: Precisely the same arguments show that $QP_3P_5P_4$ is a parallelogram with P_2, P_4, P_5 being collinear points, $P_2P_4 = P_4P_5$, and the areas $QP_2P_4P_3$ and $QP_3P_5P_4$ being equal.

We can extend this argument by induction, to show that these properties hold for all subsequent figures $QP_nP_{n+2}P_{n+1}$; and the points P_n, P_{n+2}, P_{n+3} are collinear for all n.

(vii) (a) The areas of the triangles P_n, P_{n+1}, P_{n+2} and P_{n-2}, P_n, P_{n+2} are equal for all n, and (b) each is equal to that of ΔQAB.

Proof:

(a) $\Delta P_nP_{n+1}P_{n+2}$ is half of the parallelogram $QP_nP_{n+2}P_{n+1}$; and ΔQAB is half of $QP_1P_3P_2$; the result then follows from (vi).

(b) Triangles P_n, P_{n+1}, P_{n+2} and P_{n-2}, P_n, P_{n+2} have equal areas because they have equal bases (P_nP_{n+2} and $P_{n+2}P_{n+3}$) and equal heights. □

From the above geometric observations, and theorems 2.1 and 2.2 below, it emerges that a general Fibonacci vector sequence has two interesting, fundamental properties, namely (1) the vectors $\overline{QP_n}$ tend to the upward, left and right limit rays L, L' and L'', and (2) the area of ΔQAB is an invariant property of the two types of triangle defined along any given vector sequence $\mathbf{F(a, b)}$.

It is worth noting that the geometric proofs of observations made in (v), (vi) and (vii) were elementary. Proving those properties algebraically is a more tedious exercise.

However, we must resort next to algebraic geometry, in order to obtain a general formula for the area of ΔQAB, in terms of the coordinates of $\mathbf{a}$ and $\mathbf{b}$. Fortunately we can do the job with a single vector product.

Theorem 2.1: If $\mathbf{a} = (a_1, a_2, a_3)$ and $\mathbf{b} = (b_1\, b_2, b_3)$ in $\mathbf{F(a, b)}$, then the area of ΔQAB is:

$$\Delta = \frac{1}{2}\sqrt{(a_2b_3 - a_3b_2)^2 + (a_3b_1 - a_1b_3)^2 + (a_1b_2 - a_2b_1)^2} \ .$$

From (vii), this area is constant for $\Delta P_nP_{n+1}P_{n+2}$, and for $\Delta P_{n-2}P_nP_{n+2}$, along any given Fibonacci vector sequence.

Proof: The area of $\triangle QAB$ is given by $\frac{1}{2}QA.QB\sin\angle AQB$, which is half the magnitude of the vector product $\mathbf{a} \times \mathbf{b}$. This vector product is $[(a_2b_3 - a_3b_2), (a_3b_1 - a_1b_3), (a_1b_2 - a_2b_1)]$. The formula for Δ, in the theorem, follows immediately. $\square$

We next give various results about the limit rays of vector sequences.

2.4 The Limit Rays of Vector Sequence G

In all of this sub-section, the initial vectors $\mathbf{a}$ and $\mathbf{b}$ of the vector sequence $\mathbf{G}$ are assumed to have all positive co-ordinates. Of course, if they are all negative, then the resulting limit rays will be in the same lines but with opposite senses.

The following theorem gives formulae for the direction ratios and cosines of the upward limit ray $\mathbf{L}$, of the vector sequence $\mathbf{G}$.

Theorem 2.2: The direction ratios of the limit ray $\mathbf{L}$ are $(a_1 + \alpha b_1, a_2 + \alpha b_2, a_3 + \alpha b_3)$, where α is the golden mean. Hence the direction cosines are:

$$l_i = \frac{a_i + \alpha b_i}{\sqrt{\sum_{j=1}^{3}(a_j + \alpha b_j)^2}} \, , \text{ for } i = 1, 2, 3.$$

Proof: Since $\mathbf{G}_n = F_{n-2}\mathbf{a} + F_{n-1}\mathbf{b}$, the ray from $Q(0,0,0)$ to P_n
has equation:

$$\frac{x}{F_{n-2}a_1 + F_{n-1}b_1} = \frac{y}{F_{n-2}a_2 + F_{n-1}b_2} = \frac{z}{F_{n-2}a_3 + F_{n-1}b_3} \, .$$

Hence direction ratios of ray $\overline{QP_n}$ are (dividing by F_{n-2}):

$$\left(a_1 + \frac{F_{n-1}}{F_{n-2}}b_1, \ a_2 + \frac{F_{n-1}}{F_{n-2}}b_2, \ a_3 + \frac{F_{n-1}}{F_{n-2}}b_3\right) .$$

Since $\frac{F_{n-1}}{F_{n-2}} \to \alpha$ as $n \to \infty$, and $\overline{QP_n} \to \overline{QL}$, taking limits on the bracketed coordinates gives the desired formulae for the direction ratios (r_1, r_2, r_3). The formulae for (l_1, l_2, l_3) follow immediately. $\square$

Thus, in general the limit ray of a vector sequence depends upon the initial vectors **a** and **b**. It is evident that if they are scaled, say to c**a** and c**b** respectively, the same limit ray will result, since c can then be cancelled through the direction ratios.

What is not evident, however, is that non-proportional but *different* choices of **a** and **b** can be made and *still* have the same limit ray resulting. This occurs only with choices from a very special class of initial vectors. We shall discover this class in Theorem 2.3 below.

This next theorem states the general forms of **a** and **b** which lead to the sequence's limit vector **L** being independent of the particular choices of their coordinates. The key idea is that if **a** and **b** are two consecutive terms from any given Fibonacci vector sequence, then **L** will not have any of their coordinates in its defining equations.

> **Theorem 2.3:** **L**, the limit vector of $\{G(\mathbf{a}, \mathbf{b})\}$, is the ray from Q which has equations $x/1 = y/\alpha = z/\alpha^2$, if and only if the initial vectors **a** and **b** are consecutive terms of a Fibonacci vector sequence (i.e. of *any such sequence*).
>
> **Proof:**
> (1) Let **a**, **b** be consecutive terms of a Fibonacci vector sequence. Then their forms can be written:
> $\mathbf{a} = (a_1, a_2, a_1 + a_2)$ and $\mathbf{b} = (a_2, a_1 + a_2, a_1 + 2a_2)$. Then, by Theorem 2.2, the limit vector of $\{G(\mathbf{a}, \mathbf{b})\}$ has direction ratios:
>
> $$a_1 + \alpha a_2 \quad : a_2 + \alpha(a_1 + a_2) \quad : a_1 + a_2 + \alpha(a_1 + 2a_2)$$
> i.e. $\quad a_1 + \alpha a_2 \quad : (1 + \alpha)a_2 + \alpha a_1 \quad : (1 + \alpha)(a_1 + a_2) + \alpha a_2$
> i.e. $\quad a_1 + \alpha a_2 \quad : \alpha(\alpha a_2 + a_1) \quad\quad : \alpha^2 a_1 + \alpha^3 a_2$
>
> Dividing through by $(a_1 + \alpha a_2)$, we obtain direction ratios of $1 : \alpha : \alpha^2$ for **L**. Hence the limit vector for this sequence is the ray from Q having equations $x/1 = y/\alpha = z/\alpha^2$; it is in the plane $x + y = z$. It is seen to be independent of the numbers a_1 and a_2 which serve to define **a** and **b**.

(2) Now suppose that $\mathbf{G(a,b)}$ has the limit vector as given in the theorem. Then the first direction ratio is $1 : \alpha$, and so we must have: $(a_1 + \alpha b_1) : (a_2 + \alpha b_2) :: 1 : \alpha$;

Hence $\qquad \alpha(a_1 + \alpha b_1) = \quad a_2 + \alpha b_2.$

Thus $\qquad \alpha(a_1 - b_2 + b_1) = \quad a_2 - b_1.$

Since α is irrational, this is true only if both sides vanish, in which case $b_1 = a_2$ and $b_2 = a_1 + a_2$. Also, since $\mathbf{L}$ is in plane $x + y = z$, we have $b_3 = b_1 + b_2 = a_1 + 2a_2$. Hence $\mathbf{b}$ immediately follows $\mathbf{a}$ in a Fibonacci vector sequence. $\qquad \square$

We now examine the points of Fibonacci vector sequences, to discover how they behave as n tends to $-\infty$. We shall find that there are two limiting directions in which they point away from the origin. We designate these the *left* and *right limit rays* respectively, in the obvious senses relative to the ray $\mathbf{L}$. We shall use the symbols $\mathbf{L'}$ and $\mathbf{L''}$ to denote these rays.

In order to establish formulae for the two limit rays, we first have to show that every Fibonacci vector sequence of type $\mathbf{G(a,b)}$ has a shortest vector, and that beyond that, as $n \to -\infty$ the vectors will grow longer and oscillate in direction, alternately pointing left and right.

The sequence of 'left vectors' will tend to the limit ray $\mathbf{L'}$, and the sequence of 'right vectors' will tend to the limit ray $\mathbf{L''}$. Vectors of the former type have coordinate signs $(-,+,-)$, whilst those of the latter type have coordinate signs $(+,-,+)$. Theorem 2.4 below deals with all these ideas.

There are three possible cases to be dealt with. Before stating the theorem about them, we give examples to illustrate and aid discussion. We treat only the sequence $\mathbf{G}$ having initial vectors $\mathbf{a}$ and $\mathbf{b}$ which are consecutive terms of a Fibonacci vector sequence, and for which the coordinates of $\mathbf{a}$ are all positive.

In this case we need only state the coordinate values for $\mathbf{a}$ (which we shall designate G_1), and then use the recurrence equation 'backwards' to obtain $G_0, G_{-1}, \dots$

Examples:

Case	G_1	G_0	G_{-1}	G_{-2}
(i) $a_1 > a_2$	$(4,3,7)$	$(-1,4,3)$	$(5,-1,4)$	$(-6,5,-1)$
(ii) $a_1 = a_2$	$(3,3,6)$	$(0,3,3)$	$(3,0,3)$	$(-3,3,0)$
(iii) $a_1 < a_2$	$(3,4,7)$	$(1,3,4)$	$(2,1,3)$	$(-1,2,1)$
			G_{-3}	G_{-4}
			$(11,-6,5)$	$(-17,11,-6)$
			$(6,-3,3)$	$(-9,6,-3)$
			$(3,-1,2)$	$(-4,3,-1)$

In the cases (i) and (iii) examples the shortest vectors, relative to Q, are respectively $(-1,4,3)$ and $(-1,2,1)$. Notice that each of their sign patterns is $(-,+,+)$. We show in the theorem below that this pattern must always occur in a shortest vector in these two cases, and that the shortest one is unique. Whereas in (ii) there are three shortest vectors, viz. $(0,3,3)$, $(3,0,3)$ and $(-3,3,0)$; these points are three vertices of a regular hexagon having centre Q.

Note, too, the following points:
With case (i), the shortest vector occurs immediately next to G_1; i.e. it is G_0. Case (ii) is special, in that it is a multiple of the basic Fibonacci vector sequence; and for it, one obtains three (different) shortest vectors, with the last one having the sign type $(-,+,+)$. For convenience, we shall define this last to be **the** shortest vector of the case (ii) sequence. It is easy to see that this must always occur when $a_1 = a_2$. Case (iii) needs more explanation; the number of backward steps required to arrive at the shortest vector depends on the ratio a_1/a_2 relative to the ratio $|F_{n-1}|/|F_{n-2}|$. This is so because, whether going backwards or forwards, the general term of the sequence is $F_{n-2}\mathbf{a} + F_{n-1}\mathbf{b}$.

The main points are that in each of the three cases, the shortest vector must occur as n changes from 0 to -1 to -2 etc.; it must have sign type $(-,+,+)$; and thereafter the sign types of the vectors alternate thus: $(+,-,+)$, $(-,+,-)$. We gather all this into the following theorem.

> **Theorem 2.4:** In every general Fibonacci vector sequence $\mathbf{G}(\mathbf{a}, \mathbf{b})$ with $\mathbf{a} = (a_1, a_2, a_1 + a_2)$ and both $a_1, a_2 > 0$, there is one, and only one, shortest vector, say (v_1, v_2, v_3), such that v_1 is negative, and v_2 and v_3 are both positive (v_3 is 0 in case (ii)).

The backward vector sequence proceeds, to the left of this shortest vector, with vectors which alternate in sign-type thus, $(+, -, +)$, $(-, +, -)$.

Proof:

Case (i): $(a_1 > a_2)$

Consider the three adjacent vectors G_1, G_0, G_{-1}. They are, respectively, $(a_1, a_2, a_1 + a_2)$, $(a_2 - a_1, a_1, a_2)$, $(2a_1 - a_2, a_2 - a_1, a_1)$. And in G_0, we see that $a_2 - a_1 < 0$, $a_1 > 0$ and $a_2 > 0$, therefore G_0 is of the required sign type. It is the shortest; for $|G_1| > |G_0|$ since $a_1 + a_2 > a_2 - a_1$, and $|G_{-1}| > |G_0|$ since $2a_1 - a_2 > a_2$. And as the sequence proceeds, either backwards or forwards from G_0, the lengths of the vectors increase monotonically, since in each new vector two coordinates are the same and the other is bigger (in magnitude) than the remaining one in the previous vector.

Case (ii): $(a_1 = a_2)$

Here $\mathbf{a} = (a_1, a_1, 2a_1) = a_1(1, 1, 2)$; and $\mathbf{b} = a_1(1, 2, 3)$. Hence $\mathbf{G}(\mathbf{a}, \mathbf{b})$ is equal to $a_1\mathbf{F}$, a multiple of the basic Fibonacci vector sequence. Since $\mathbf{F}$ has three shortest vectors, viz. $\mathbf{F}_{-1} = (-1, 1, 0)$, $\mathbf{F}_0 = (1, 0, 1)$ and $\mathbf{F}_1 = (0, 1, 1)$, then so does $\mathbf{G}$. The unique one required for the theorem is $a_1\mathbf{F}_{-1}$.

Case (iii): $(a_1 < a_2)$

Consider what happens when passing from $G_1 = (a_1, a_2, a_1 + a_2)$ to $G_0 = (a_2, a_1, a_2)$. Since $a_2 - a_1 > 0$, the signs in G_0 are $(+, +, +)$. But in G_0, it may be that $a_2 - a_1 > a_1$; in which case G_0 will be a vector of Case (i) type: so a second 'step backwards' to G_{-1} will take us to the shortest vector. For this, the requirement on G_0 is that $a_2 > 2a_1$, that is $a_2/a_1 > 2(= F_3/F_2)$, note).

Example: $G_1(3, 7, 10) \rightarrow G_0(4, 3, 7) \rightarrow G_{-1}(-1, 4, 3)$, the shortest vector. Note that the coordinate signs are $(-, +, +)$.

If this requirement is not met, we must take another step backwards, to $G_{-1} = (2a_1 - a_2, a_2 - a_1, a_1)$. The sign-type of this triple is $(+, +, +)$, since $2a_1 - a_1 > 0$, $a_2/a_1 < 2$ and $a_2 - a_1 > 0$. Hence we must step backwards again, to $G_{-2} = (2a_2 - 3a_1, 2a_1 - a_2, a_2 - a_1)$.

This will be of the required sign-type $(-, +, +)$ if the first coordinate is negative. That is, if $a_2/a_1 < 3/2 (= F_4/F_3)$. [An example starts from $G_2 = (3, 4, 7)$, shown in the table above.]

As this process continues, we have to compare the ratio a_2/a_1 with F_n/F_{n-1}, using alternately the relations $<$ and $>$. Only if the ratio a_2/a_1 lies forever between alternate values of F_n/F_{n-1} will the process never cease. This cannot happen,

for $F_n/F_{n-1} \to \alpha$, which is irrational; eventually one of the inequalities must be satisfied; and then a vector of sign-type $(-, +, +)$ will have been reached. By the reasoning used at the end of Case (i), this must be the shortest vector in the sequence.

To complete the proof, we have to show that after the shortest vector, the backward sequence proceeds with vectors which alternate in sign types thus: $(+, -, +), (-, +, -), \cdots$ This follows immediately from the way in which each new vector is formed from the current one. $\qquad \square$

We are now in a position to obtain formulae for the two limit rays, $\mathbf{L}'$ and $\mathbf{L}''$, which obtain when $n \to -\infty$. From the proof of Theorem 2.2, we know that the vectors in the sequence $\mathbf{G}$ have direction ratios thus:

$$\left(a_1 + \frac{F_{n-1}}{F_{n-2}} b_1\right) : \left(a_2 + \frac{F_{n-1}}{F_{n-2}} b_2\right) : \left(a_3 + \frac{F_{n-1}}{F_{n-2}} b_3\right).$$

When $n < 0$, F_{n-1} and F_{n-2} differ in sign, and $|F_{n-1}| < |F_{n-2}|$. With this knowledge, taking limits as $n \to \infty$ gives the following direction ratios for the limit rays:

$$r_1 : r_2 : r_3 = \left(a_1 - \frac{1}{\alpha} b_1\right) : \left(a_2 - \frac{1}{\alpha} b_2\right) : \left(a_3 - \frac{1}{\alpha} b_3\right).$$

These determine a line through the origin in the plane of $\mathbf{a}$ and $\mathbf{b}$, which has equation $x/r_1 = y/r_2 = z/r_3$.

$\mathbf{L}'$ is the limit ray for those points (alternating in $\mathbf{G}$) which move away from the origin in the direction of this line to the left of $\mathbf{L}$; and $\mathbf{L}''$ is the limit ray for the other points of $\mathbf{G}$ moving in the opposite direction : the union of $\mathbf{L}'$ and $\mathbf{L}''$ is the line itself.

It should be noted that **all** points of a given sequence **G** lie above (or all below) this line.

To give but one example, we use again the case of Theorem 2.3, taking $\mathbf{a} = (a_1, a_2, a_3)$ and $\mathbf{b} = (a_2, a_1 + a_2, a_1 + 2a_2)$.

The combined limit line has direction ratios:

$$(a_1 - \frac{1}{\alpha}a_2) \, : \, (a_2 - \frac{1}{\alpha}(a_1 + a_2)) \, : \, ((a_1 + a_2) - \frac{1}{\alpha}(a_1 + 2a_2)) \,.$$

This gives the direction for determining the two limit rays through the origin $Q(0,0,0)$. One is for the sequences of odd-negative-subscripted terms of **all** the Fibonacci vector sequences; and the other, in the opposite direction, is for the even-negative-subscripted terms from those sequences.

The direction ratios are independent of the values of a_1 and a_2; that is why the same limit rays apply to all Fibonacci vector sequences in the plane determined by this **a** and **b**.

Continuation: We shall continue this study of limit rays, and make further geometric observations on the vector sequences, in Chapter 6, after having had the opportunity to study more deeply the generation of vector sequences in planes.

The next three chapters concentrate on geometry in the honeycomb plane $x + y - z = 0$, and the Fibonacci and Lucas vector polygons.

Chapter 3

The Fibonacci Honeycomb Plane

In this chapter we shall study the locations, and various configurations, of the integer-points (those with integer coordinates) in the plane $x+y-z = 0$, wherein lie all the Fibonacci vectors. It will soon become apparent why we have named this plane *the honeycomb plane*. We shall often refer to it by the symbols π_0.

In particular, properties of points of the Fibonacci and Lucas vector sequences, and geometric facts related to them, will be discovered.

3.1 A Partition of the Integer Lattice

Let $(a, b, c) \in \mathbf{Z}^3$ be a point in the integer lattice $\mathcal{L}$. We can partition the points of $\mathcal{L}$ into three non-intersecting sets thus:

$$\begin{aligned} S_1 &= \{(a,b,c) : a+b > c\} \\ S_2 &= \{(a,b,c) : a+b = c\} \\ S_3 &= \{(a,b,c) : a+b < c\} \end{aligned}$$

Note that, if (a, b, c) is in the positive octant and represents lengths of three sides of a triangle, then S_1, S_2 and S_3 represent, respectively, the sets of 'real', 'degenerate' and 'virtual' integer triangles.

> **Theorem 3.1:** Points in S_2 lie in the plane $x + y - z = 0$; call this plane π_0. S_1 is the set of points in $\mathcal{L}$ which lie below π_0; whereas S_3 is the set of points in $\mathcal{L}$ which lie above π_0.

We turn now to a study of how Fibonacci vectors and Fibonacci vector sequences behave in plane π_0.

3.2   Fibonacci Vectors in the Plane π_0

We first examine how the points in π_0 determine Fibonacci vectors and Fibonacci vector sequences.

> **Theorem 3.2:**   Any point (a, b, c) in π_0 determines the general Fibonacci sequence $F(a, b)$, since $c = a + b$ (see definitions (1.1) and (1.2) above). Hence it is a Fibonacci vector G_n. If a and b are integers, we may call these *Fibonacci points*.

> **Theorem 3.3:**   Three different *vector sequences* may be distinguished within $F(a, b)$, each of which contains amongst the vectors' coordinates all elements of $F(a, b)$ exactly once. These are:

$$G_{(1)} = \{\ldots, (a, b, a + b), (a + 2b, 2a + 3b, 3a + 5b), \ldots\}$$
$$G_{(2)} = \{\ldots, (b, a + b, a + 2b), (2a + 3b, 3a + 5b,), \ldots\}$$
$$G_{(3)} = \{\ldots, (a + b, a + 2b, 2a + 3b), (3a + 5b, 5a + 8b, 8a + 13b), \ldots\}$$

Proofs are not needed for the above two theorems: each follows directly from definitions.

Note that if (a, b, c) is the point $Q(0, 0, 0)$, these vector sequences are identically equal to the constant vector sequence $\{\ldots, (0, 0, 0), (0, 0, 0), \ldots\}$, which is Q repeated indefinitely.

We shall denote the union of these three sets of vectors by $\mathbf{G} = \{\mathbf{G}_n\}$, and call it the *general Fibonacci vector sequence*.

Examples:

(1) The *basic Fibonacci vector sequence* $\mathbf{F}$ is defined to be the union of the following three vector sequences:

$$\mathbf{F}_{(1)} = \{\mathbf{F}_{-1}\} = \{\ldots, (-1, 1, 0), (1, 1, 2), (3, 5, 8), (13, 21, 34), \ldots\}$$
$$\mathbf{F}_{(2)} = \{\mathbf{F}_0\} = \{\ldots, (1, 0, 1), (1, 2, 3), (5, 8, 13), (21, 34, 55), \ldots\}$$
$$\mathbf{F}_{(3)} = \{\mathbf{F}_1\} = \{\ldots, (2, -1, 1), (0, 1, 1), (2, 3, 5), (8, 13, 21), \ldots\}$$

Thus: $\mathbf{F} = \{\mathbf{F}_n\} = \{\mathbf{F}_{(1)}\} \cup \{\mathbf{F}_{(2)}\} \cup \{\mathbf{F}_{(3)}\}$. We shall usually omit the word 'basic', when referring to this Fibonacci vector sequence, if the context allows it.

It will be seen that the three subsequences of $\mathbf{F}$, listed above, are obtained from the basic sequence by taking every third Fibonacci vector in turn, and placing it into its appropriate subsequence. Symbolically, this appears as follows:

$$\begin{aligned}
\mathbf{F}_{(1)} &= \{..., \mathbf{F}_{-1}, \mathbf{F}_2, \mathbf{F}_5, ... \} \\
\mathbf{F}_{(2)} &= \{..., \mathbf{F}_0, \mathbf{F}_3, \mathbf{F}_6, ... \} \\
\mathbf{F}_{(3)} &= \{..., \mathbf{F}_1, \mathbf{F}_4, \mathbf{F}_7, ... \}
\end{aligned}$$

A glance at the actual numbers (coordinates) in the vector sequences written out in detail above, shows that the following vector recurrence equation will generate each of the subsequences of vectors.

$$\mathbf{S}_{m+2} = 4\mathbf{S}_{m+1} + \mathbf{S}_m .$$

A proof of this recurrence, using elementary Fibonacci identities, is simple. Since each vector, in all three subsequences, is a Fibonacci vector, all subsequence vectors approach the same limiting ray as n tends to infinity, namely the line L_∞ from Q, which has direction-ratios $1 : \alpha : \alpha^2$.

It is obvious how we may generalise the above procedure; for example, if we took every fourth vector from the Fibonacci vector sequence, we should obtain four subsequences whose union would be $\mathbf{F}$. This time, however, we would not have the attractive property that each subsequence uses all the Fibonacci numbers exactly once amongst its coordinates; every fourth one would be missing. For example, the vector subsequence which includes $\mathbf{F}_0$ is ..., $(1, 0, 1)$, $(2, 3, 5)$, $(13, 21, 34)$, ... The 'gaps' in the Fibonacci number sequence are $1, 8, 55, ...$ etc. However, we may note that the differences between successive vectors in the subsequence are Lucas vectors. And that the subsequences of vectors are now generated by the following vector recurrence:

$$\mathbf{S}_{m+2} = 7\mathbf{S}_{m+1} - \mathbf{S}_m .$$

If we generalize this process still further, we can take every kth vector from the basic Fibonacci vector sequence, and obtain a union of k vector subsequences for $\mathbf{F}$. These subsequences are generated by the vector recurrence (where L_k is the kth Lucas number):

$$\mathbf{S}_{m+2} = L_k \mathbf{S}_{m+1} + (-1)^{k-1} \mathbf{S}_m .$$

(2) For a second numerical example, we show the Lucas vector sequence, separated into a union of three subsequences, as was done with the basic Fibonacci vector sequence.

$$\{\mathbf{L}_{(1)}\} \;=\; \{\mathbf{L}_{-1}\} \;=\; \{\ldots, (-11,7,-4), (3,-1,2), (1,3,4), (7,11,18),\ldots\}$$
$$\{\mathbf{L}_{(2)}\} \;=\; \{\mathbf{L}_{0}\} \;=\; \{\ldots, (7,-4,3), (-1,2,1), (3,4,7), (11,18,29),\ldots\}$$
$$\{\mathbf{L}_{(3)}\} \;=\; \{\mathbf{L}_{1}\} \;=\; \{\ldots, (-4,3,-1), (2,1,3), (4,7,11), (18,29,47),\ldots\}$$

Thus: $\mathbf{L} = \{\mathbf{L}_n\} = \{\mathbf{L}_{(1)}\} \cup \{\mathbf{L}_{(2)}\} \cup \{\mathbf{L}_{(3)}\}$.

For these three vector subsequences, the generating recurrence is again $\mathbf{S}_{m+2} = 4\mathbf{S}_{m+1} + \mathbf{S}_m$. And the limit vector is again the ray through Q with direction ratios $1 : \alpha : \alpha^2$.

> **Theorem 3.4:**   Every integer point in π_0 belongs to one and only one general Fibonacci vector sequence.
>
> **Proof:**   Given a point (a,b,c) in π_0. Since $c = a + b$ the point determines uniquely some general Fibonacci vector sequence, which contains the vector $(a, b, a + b)$. Hence that point cannot occur in any other general Fibonacci vector sequence. $\square$
>
> **Corollary:** The set of integer points of the plane π_0 is partitioned by the class of all general Fibonacci vector sequences.
>
> We remark here that we do not need to confine ourselves to integer points. For example, we could take the point $(1, \alpha, \alpha^2)$. It lies in the plane π_0; and its associated Fibonacci sequence is $1, \alpha, \alpha^2, \alpha^3, \ldots$. This is a geometric progression with common ratio α; but it is also a Fibonacci sequence, since $\alpha^{n+2} = \alpha^{n+1} + \alpha^n$. Moreover, every one of the vectors derived from it, taking three consecutive members as coordinates, has the same direction ratios, namely $1 : \alpha : \alpha^2$. Thus they all lie upon the limit vector L_∞ that we have talked much about above. Indeed, if we join the vector-points up, from this sequence, the resulting line segments constitute L_∞! It is highly fitting that the geometric sequence of the Golden Mean should do this.

We move now from our discussion on vector sequences in the Fibonacci plane, in order to study how points in the plane are situated. We shall return to the vector sequences later, when we know more about the arrangement of the integer points in π_0.

3.3   The Honeycomb of Points in π_0

We shall now show that the points in the plane π_0 form a honeycomb arrangement, which we shall call *The Fibonacci Honeycomb*.

First we show that every integer point P in π_0 has six neighbouring points in π_0 placed symmetrically with respect to it. They determine a regular hexagon having P as centre. It will follow from this, that the set of points in π_0 can be covered by three sets of parallel lines, arranged at $60°$ to one another.

Nearest neighbours of $P(a, b, a + b)$ are:

$$
\begin{array}{rcl}
P_1 & \equiv & (a, b+1, a+b+1) \\
P_2 & \equiv & (a+1, b, a+b+1) \\
P_3 & \equiv & (a+1, b-1, a+b) \\
P_4 & \equiv & (a, b-1, a+b-1) \\
P_5 & \equiv & (a-1, b, a+b-1) \\
P_6 & \equiv & (a-1, b+1, a+b)
\end{array}
$$

Theorem 3.5 below shows that these points determine a regular hexagon centred at $P(a, b, a + b)$.

> **Theorem 3.5:**
> Let $P = (a, b, c)$ be any integer point in π_0. Then its nearest neighbours are the six points P_i defined above. They are arranged in a regular hexagon about P, of side length $\sqrt{2}$ and diameter $2\sqrt{2}$.
>
> **Proof:** It may quickly be established that the Euclidean distances PP_i, with $i = 1, \ldots, 6$, and also the distances $P_i P_{i+1}$, with $i = 1, \ldots, 5$ and $P_6 P_1$ are all equal to $\sqrt{2}$. In view of the relation $a + b = c$ for points in π_0, all other points are further than $\sqrt{2}$ from P, since all possible arrangements where a 1 is added to or subtracted from the coordinates of P have been included amongst the six points; adding or subtracting a 2 (or more) to any coordinate will require a 2 (or more) to be added or subtracted to one of the other coordinates; then the distance from P to this new point will be greater than $\sqrt{2}$.
> $\square$

The resulting hexagon is shown as the left-hand diagram of Fig. 1. The right-hand diagram shows the hexagon having $Q(0,0,0)$ as centre-point.

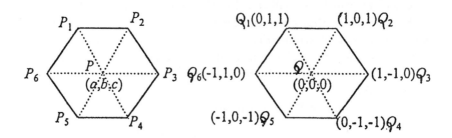

The neighbours of $P(a,b,c)$ *The neighbours of $Q(0,0,0)$*

Figure 1. Hexagons in the plane π_0.

Theorem 3.6: Let $Q(0,0,0)$ be designated the centre (origin) of the plane π_0. Let l,m,n be symbols for three sets of parallel lines, which are defined as follows:

l_0 is the line through Q and $Q_1(0,1,1)$; it has direction cosines $(0,1,1)/\sqrt{2}$.

l is the set of all lines parallel to l_0, separated by perpendicular distances which are multiples of $\sqrt{6}/2$.

m_0 is the line through Q and $Q_2(1,0,1)$; it has direction cosines $(1,0,1)/\sqrt{2}$.

m is the set of all lines parallel to m_0, separated by perpendicular distances which are multiples of $\sqrt{6}/2$.

n_0 is the line through Q and $Q_3(1,-1,0)$; it has direction cosines $(1,-1,0)/\sqrt{2}$.

n is the set of all lines parallel to l_0, separated by perpendicular distances which are multiples of $\sqrt{6}/2$.

Any three lines, one of each of l,m, and n, intersect in a point. Each of the sets of parallel lines (i.e. each of l,m,n) covers every integer point of the plane π_0.

Proof: This is immediately evident, as a glance at Fig. 2 will confirm. The sets of lines l,m,n triangulate the plane π_0, into equilateral triangles of side $\sqrt{2}$ and height $\sqrt{6}/2$.

The set of vertices of all the triangles constitutes the set of integer points in π_0.

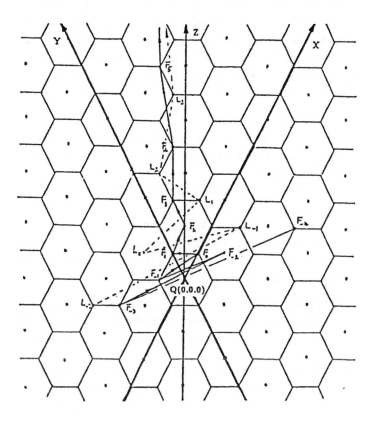

Figure 2. The Fibonacci Honeycomb—the plane π_0 and axes

The centres of the hexagons are the B-points;
all other integer points are H-points.
Also shown are the Fibonacci and Lucas vector polygons.

Definition 3.1: We shall call m_0 the X-axis, l_0 the Y-axis; and the Z-axis is taken to be the line through Q (in π_0) which is perpendicular to n_0.
We shall use the set of these three axes, with $Q(0,0,0)$ as origin, as a reference frame, in order to define points and figures in π_0.

Unit 'steps' along both the X- and Y-axes are to be of length $\sqrt{2}$, whereas unit 'steps' along the Z-axis are to be of length $\sqrt{6}/2$. With these conventions, taking $X = a$ and $Y = b$ (with a, b being integers) will serve to define the integer point $P(a, b, c)$. Note that actually the Z-axis is redundant, since in π_0 we have $c = a + b$; it is often convenient to use it, however. Finally, on occasion we shall refer to the line through Q which is perpendicular to the Z-axis as the Z'−axis.

If we take the hexagon drawn on the nearest neighbours of Q, it is evident that we can draw six hexagons around it, on its sides; then continue drawing hexagons on the outer sides, and so on, until the whole plane is tessellated with hexagons.

It will follow that by this tessellation we partition the points of π_0 into two sets, namely:

$B \equiv$ the set of points at the hexagon centres;

$H \equiv$ the set of points on all the hexagons.

We may call the points in B the B-points (Bees!) of π_0; and the points in H the H-points (Honey points).

> **Definition 3.2:** The H-points and the lines parallel to the three axes which join them constitute *the Fibonacci Honeycomb*, with the B-point $Q(0, 0, 0)$ being *the Queen Bee*. (see Fig. 2)

Before embarking on a study of Fibonacci vector sequences and other geometric figures on the Honeycomb, we shall state one or two useful theorems about B- and H-points. A great deal more can be said about these sets of points, and lines joining them, than we have space for here.

3.4   Some Properties of B-points

This section gives a method for determining whether a point in Pi_0 is a B-point. Then it treats several properties of the B-points. **Location of the B-points**

Let $P(a, b, c)$ be any point in π_0. The following algorithm determines whether or not P is a B-point. (Note that $|d|_3$ designates the remainder from d after division by 3; i.e. 'remainder modulo 3'.)

Algorithm: Compute $|a|_3$ and $|b|_3$. Then P is a B-point if and only if both these remainders modulo 3 are equal.

Proof: We can arrive at P in two moves, travelling from Q to P as follows:

Move (i): Move a hexagon-side lengths (i.e. $a\sqrt{2}$) in the direction QQ_2 (along the X-axis); move up if a is positive, and down if a is negative.

Move (ii): Move b hexagon-side lengths (i.e. $b\sqrt{2}$) in the direction QQ_1 (parallel to the Y-axis); move up if b is positive, and down if b is negative.

There are three possibilities to be considered, for move (i) followed by move (ii). They are:

First: If $|a|_3 = 0$, then move (i) arrives at a B-point, after which move (ii) will lead to a B-point if and only if $|b|_3 = 0$, because of the honeycomb structure.

Second: If $|a|_3 = 1$, then move (i) arrives at an H-point, after which move (ii) will lead to a B-point if and only if $|b|_3 = 1$, because of the honeycomb structure.

Third: If $|a|_3 = 2$, then move (i) arrives at an H-point, after which move (ii) will lead to a B-point if and only if $|b|_3 = 2$, because of the honeycomb structure.

Since these are the only possible ways in which a B-point can be arrived at, the Theorem is proved.

Evidently, all other points in π_0 are H-points. $\square$

Corollary 1: It easily follows from the theorem that P is a B-point iff $|a + c|_3 = 0$.

Corollary 2: The above algorithm shows that we can classify the B-points into three types, as follows:

$$A\ B\text{-point}\ P(a,b,c)\ \text{is of type} \begin{cases} B_0, & \text{iff } |a|_3 = 0\ ; \\ B_1, & \text{iff } |a|_3 = 1\ ; \\ B_2, & \text{iff } |a|_3 = 2\ . \end{cases}$$

Corollary 3: The set of B-points in π_0 which are of type B_0 form an Abelian group under vector addition; the identity element is the Queen Bee (i.e. $Q(0,0,0)$!).

Proof: Q is a point of type B_0. Take any two points of type B_0 and add them. In the resulting point, the value of $|a + c|_3$ is zero; so by Corollary 1, the new point is also of type B_0. Hence the set is closed under addition; and vector addition is commutative; so the set forms an Abelian group under addition. $\square$

Corollary 4: Addition of B-point types can be defined in the obvious way, since the addition of elements of two given types always leads to an element of the same type.

Below we give the addition table for B-point types. It shows that the three types form an Abelian group under addition, with the identity element being the type B_0. This group is the cyclic group of order 3.

$+$	B_0	B_1	B_2
B_0	B_0	B_1	B_2
B_1	B_1	B_2	B_0
B_2	B_2	B_0	B_1

Much more can be said about points and lines in π_0, and their geometry, but space does not allow it. Triangles are studied in Chapter 5.

We shall close the chapter by stating and proving a theorem about line segments in the Fibonacci plane. The objective is to show that no integer-segment (that is, no segment whose endpoints are integer-vectors) in this plane has integer length.

Theorem 3.7: Let $A = (a_1, a_2, a_3)$ and $B = (b_1, b_2, b_3)$, with all six coordinates being integers, be any two integer-points in the plane π_0. Denote by AB the line-segment joining them; and by $|AB|$ the length of this segment.
Then $|AB| = \sqrt{[(a_1 - b_1)^2 + (a_2 - b_2)^2 + (a_3 - b_3)^2]}$ is not integral.

Proof: Let $A'B'$ and $A''B''$ be the projections of AB onto QZ and QZ' respectively (see Fig. 3). They have lengths

ma, and nb respectively, where $a = \sqrt{6}/2$, $b = \sqrt{2}/2$, and m, n are integers (the projections are from integer-points).

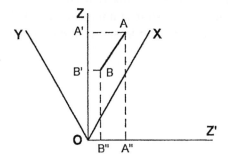

Figure 3. Projections of AB onto the z-axes.

Hence: $|AB|^2 = m^2 a^2 + n^2 b^2 = m^2 \dfrac{6}{4} + n^2 \dfrac{1}{2}$.

Suppose that $|AB|$ is an integer d. Then we have to show that this is impossible; namely, that the equation $3m^2 + n^2 = 2d^2$ has no integer triple (m, n, d) solution.

We begin by assuming that there is an integer solution, with $gcd(m, n, d) = 1$, and then derive a contradiction.

First observe that the equation implies that either (i) both m and n are even, or (ii) both m and n are odd.

In case (i), set $m = 2m'$ and $n = 2n'$, and note that d must be odd, since $\gcd(m, n, d) = 1$. The equation is then:
$$2d^2 = 12m'^2 + 4n'^2 \rightarrow d \text{ is even}.$$

This is a contradiction, therefore (i) is impossible.

In case (ii), we shall prove that $3m^2 = 2d^2 - n^2$, with m, n both odd, has no integer solution.

The right-hand side must be divisible by 3.
Assume that 3 divides both d and n ; then 3 doesn't divide m, since $\gcd(m, n, d) = 1$. Put $d = 3r$ and $n = 3s$ and the equation becomes: $m^2 = 6r^2 - 3s^2$ which $\rightarrow 3|m$. This is a contradiction, hence the assumption is wrong.

Then, for 3 to divide evenly the right-hand side of the equation, the remainders modulo 3 of $2d^2$ and $-n^2$ must add to 3. But it is easy to show that both $2d^2$ and $-n^2$ are either 0 or 2 modulo 3; so it is impossible for their remainders to sum to three, when at least one of d or n is not a multiple of 3. □

Chapter 4

Fibonacci and Lucas Vector Polygons

Recall that the Fibonacci vector sequence $\{\mathbf{F}_n\}$ is:
$$\{\ldots, (2, -1, 1), (-1, 1, 0), (1, 0, 1), (0, 1, 1), (1, 1, 2), (1, 2, 3), (2, 3, 5), \ldots\},$$
with the general element being $\mathbf{F}_n \equiv (F_{n-1}, F_n, F_{n+1})$.

In similar fashion, we can write down the Lucas vector sequence $\{\mathbf{L}_n\}$.

Definition 4.1: If we join the 'points' of the Fibonacci vector sequence, plotted in the Honeycomb plane, by straight lines, we obtain a geometric figure which we shall call the *Fibonacci Polygon*. Similarly, we can join the points of the Lucas vector sequence, to obtain the *Lucas Polygon*. Fig. 2 (in Chap. 3) shows these two polygons drawn in the Honeycomb plane.

4.1 Convergence Properties of the Polygons

It may be observed that both polygons progress indefinitely upwards, and appear to tend towards a line which passes through Q and has a direction which is to the left of the Z-axis. Below, we find the equation of this line.

It may also be observed that there are **two other limit directions** for the polygons; the vectors $\mathbf{F}_n$ with negative and odd values of n tend to one of these, whilst those with negative even values of n tend to the other.

We shall use the notations $\mathbf{u}, \mathbf{v}, \mathbf{w}$ for the direction cosines of these three limit lines, respectively.

Theorem 4.1: The direction cosines $\mathbf{u}$ are $(1/2\alpha, 1/2, \alpha/2)$, where α is the golden ratio.

Proof: Since Q is the point $(0,0,0)$, and $\mathbf{F}_n$ is the point (F_{n-1}, F_n, F_{n+1}), the direction cosines of the vector QP_n are: $(F_{n-1}, F_n, F_{n+1})/\sqrt{(F_{n-1}^2 + F_n^2 + F_{n+1}^2)}$.

Dividing numerator and denominator by F_{n-1}, and letting n tend to infinity, for each coordinate, we obtain the required result. $\qquad\qquad\qquad\qquad\qquad\qquad\qquad\qquad\Box$

In like manner, we can take the general Fibonacci vector sequence, derived from $F(a,b)$, and compute the limit of vector QP_n. We find that the result is independent of the choice of a and b (except that at least one must be nonzero). This means that all (except $\mathbf{F}(0,0)$) Fibonacci vector sequences tend to the same ray QP_∞ which has direction cosines $\mathbf{u}$. The equations of this line are: $\alpha x = y = z/\alpha$.

Theorem 4.2: The direction cosines for $\mathbf{v}$ and $\mathbf{w}$ are, respectively,

$$(\alpha/2, -1/2, 1/2\alpha) \text{ and } (-\alpha/2, 1/2, -1/2\alpha).$$

Proof: The proof follows similar lines to that for Theorem 4.1. It will be left to the reader.

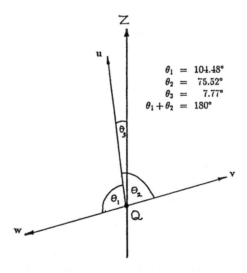

$$\theta_1 = 104.48°$$
$$\theta_2 = 75.52°$$
$$\theta_3 = 7.77°$$
$$\theta_1 + \theta_2 = 180°$$

Figure 1. The limit lines in π_0, for the Fibonacci polygons.

The diagram of Fig. 1 above summarizes these results, and shows the angles between the limit lines. Note that **v** and **w** are collinear: $\mathbf{v}.\mathbf{w} = -1$. The line $x = -\alpha y = \alpha^2 z$ is a lower bound to both polygons.

4.2 Some Theorems about Triangles, Lines, and Quadrilaterals

The following results are about triangles, lines and quadrilaterals which are drawn upon Fibonacci vector polygons.

Theorem 4.3: Let **U**, **V**, and **W** be three consecutive points in a general Fibonacci polygon. Let $\mathbf{U} = (a, b, a + b)$. Consider their relationship with $Q(0,0,0)$; in particular, consider the quadrilateral $QUWV$ (see Fig. 2). Then:

(i) The figure $QUWV$ is a parallelogram.

(ii) $\Delta(QUV) = \Delta(UVW) = \frac{\sqrt{3}}{2}|b^2 - a^2 - ab|$.

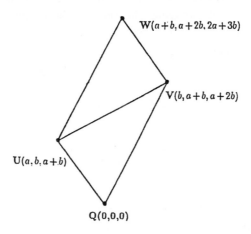

Figure 2. The quadrilateral QUWV defined in Theorem 4.3.

Proof:

(i) Since $\mathbf{U} - \mathbf{Q} = \mathbf{W} - \mathbf{V}$ we have $QU \parallel VW$. Similarly, $QV \parallel UW$. Hence $QUWV$ is a parallelogram; and $\Delta(QUV) = \Delta(UVW)$.

(ii) The area of $\Delta(QUV)$ is given by $\sqrt{(\Delta_1^2 + \Delta_2^2 + \Delta_3^2)}$ where:

$$2\Delta_1 = \begin{Vmatrix} 0 & 0 & 1 \\ a & b & 1 \\ b & a+b & 1 \end{Vmatrix} = |b^2 - a^2 - ab|. \text{ Similarly}$$

we find that both $2\Delta_2$ and $2\Delta_3$ are equal to $|b^2 - a^2 - ab|$. Hence the theorem's result for Δ.   □

Theorem 4.4: For any given Fibonacci vector sequence, the triangle formed by any three consecutive Fibonacci vectors has an area which is constant throughout the sequence. The formula for this area is $\frac{\sqrt{3}}{2}|b^2 - a^2 - ab|$.

Proof: Take two consecutive general Fibonacci vectors, namely $U(F_{n-3}a+F_{n-2}b,\ F_{n-2}a+F_{n-1}b,\ F_{n-1}a+F_nb)$ and $V(F_{n-2}a+F_{n-1}b,\ F_{n-1}a+F_nb,\ F_na+F_{n+1}b)$, and then compute the area of triangle QUV. It is found that, after reduction using Fibonacci identities, the area is the formula of the theorem. Then, applying Theorem 4.3 shows that it is constant throughout the sequence.   □

Corollaries:
(i) The area of $\Delta(\mathbf{F}_{n-1}\mathbf{F}_n\mathbf{F}_{n+1})$ is constant, and equals $\sqrt{3}/2$.
(ii) The area of $\Delta(\mathbf{L}_{n-1}\mathbf{L}_n\mathbf{L}_{n+1})$ is constant, and equals $5\sqrt{3}/2$.

The final three theorems give interesting geometric results about the Fibonacci and Lucas polygons. They will be stated, and left to the reader to prove.

Theorem 4.5: The triangle $\Delta(\mathbf{L}_{n-1}, \mathbf{L}_n, \mathbf{L}_{n+1})$ includes the point $\mathbf{F}_{n+2}$. In fact, we have that:
$\mathbf{F}_{n+2} = \frac{1}{5}\mathbf{L}_{n-1} + \frac{2}{5}\mathbf{L}_n + \frac{2}{5}\mathbf{L}_{n+1}$, for all n.

Theorem 4.6: Side $\mathbf{L}_n\mathbf{L}_{n+1}$ cuts side $\mathbf{F}_{n+2}\mathbf{F}_{n+3}$ in the ratio $\frac{1}{3}:\frac{2}{3}$ internally; and vice versa. It also cuts side $\mathbf{F}_{n+1}\mathbf{F}_{n+2}$ in the ratio $2:1$ externally.

Theorem 4.7: The quadrilateral $\mathbf{L}_n\mathbf{F}_{n+1}\mathbf{L}_{n+1}\mathbf{F}_{n+3}$ is a parallelogram, for all n, of constant area $2\sqrt{3}$.

Chapter 5

Trigonometry in the Honeycomb Plane

In this chapter, we present some theorems about triangles which occur in the Fibonacci plane π_0. We shall be concerned only with triangles on the lattice points; we shall call these *integer triangles*.

We have already given some results on triangles related to Fibonacci vector polygons, in the previous chapter. In this chapter, we begin by showing that a $(90°, 45°, 45°)$ integer triangle is impossible in π_0. By contrast, integer equilateral triangles are 'everywhere', which we demonstrate by giving a general formula for computing equilateral triangles in π_0, and develop some construction algorithms using the formula. Finally we shall define and study transforms of triangles by means of equilateral triangle constructions, and apply them to Fibonacci vector polygons.

5.1 The Impossibility of a (90,45,45) Integer Triangle

From Definition 3.1 in Chapter 3, it is easy to see how the projections of a lattice segment in π_0 onto the z-axes QZ and QZ' are integer multiples of $\sqrt{6}/2$ and $\sqrt{2}/2$ and respectively. We may use this fact to prove that a $(90°, 45°, 45°)$ integer triangle cannot exist in the honeycomb plane π_0.

> **Theorem 5.1:** Let ABC be a triangle in the honeycomb plane $x + y = z$, having angles $90°, 45°, 45°$. Then not all of the vertices A, B, C are integer points; i.e. the triangle is not an integer triangle.

Proof: $\triangle ABC$ is isosceles, with $AB = AC$. If we project the arm AB onto the axis QZ, and the arm AC onto the axis QZ', then the two projections will be of equal length (since $AB \perp BC$ and $QZ \perp QZ'$, the projection angles are equal).

Suppose that each vertex is an integer point. Then, since an integer point projects into an integer point on an axis, there must exist integers, say m and n, such that $m\sqrt{6}/2 = n\sqrt{2}/2$. But that means that $n = m\sqrt{3}$, which is impossible, since $\sqrt{3}$ is irrational. Hence a contradiction, therefore the theorem is true.   □

Corollary: It follows immediately that no square with integer vertices exists in the honeycomb plane.

5.2 The Ubiquity of (60,60,60) Integer Triangles

We now will show in what sense equilateral triangles with integer vertices are 'everywhere' in the honeycomb plane.

Let $A(a_1, a_2, a_3)$ and $B(b_1, b_2, b_3)$ be any two lattice points in π_0. Then we know that $a_3 = a_1 + a_2$ and $b_3 = b_1 + b_2$.

Consider the following diagram, which shows equilateral triangles drawn on either side of AB, with ABL anticlockwise and ABR clockwise.

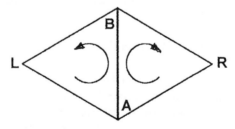

We shall say that point L is to the left of AB (directed); and point R is to the right of AB. We can conveniently write $L = E(AB)$ and $R = E(BA)$, using the single letter E to mean 'equilateral triangle transform'. We shall sometimes call these the ET-transforms of AB. Note that both $\triangle ABL$ and $\triangle BAR$ are traversed in the anticlockwise direction.

Theorem 5.2: With A and B defined as above, the ET-transforms of AB are given by the following formulae:

$$L = (a_1 + a_2 - b_2, b_1 + b_2 - a_1, a_2 + b_1)$$
$$R = (b_1 + b_2 - a_2, a_1 + a_2 - b_1, a_1 + b_2)$$

Clearly both L and R are lattice points.

Proof: (i) For $\triangle ABL$, we have:

$$A - B = (a_1 - b_1, a_2 - b_2, a_1 + a_2 - b_1 - b_2)$$
$$B - L = (b_1 + b_2 - a_1 - a_2, a_1 - b_1, b_2 - a_2)$$
$$L - A = (a_2 - b_2, a_1 + a_2 - b_1 - b_2, b_1 - a_1)$$

These are the vectors for the three sides of the triangle. The squared-lengths of the sides are obtained by summing the squares of the coordinates of each of these vectors in turn. It is easy to see that these sums are all equal; hence the three sides are equal, so $\triangle ABL$ is equilateral.

(ii) Similar calculations for $A - B$, $B - R$ and $R - A$ show that $\triangle ABR$ is also equilateral. $\square$

Hence every pair of lattice points in π_0 is a side of two equilateral triangles, with all vertices being lattice points.

Theorem 5.3: If A and B are B-points in π_0, then so are both L and R.

Proof: We showed in Chapter 3 (section 3.4) that (b_1, b_2, b_3) is a B-point iff $b_1 + b_3 \equiv 0 \pmod 3$.

From Theorem 5.2, the formulae for L and R give:

$$(i) \quad l_1 + l_3 = a_1 + 2a_2 + b_1 - b_2 \,;$$
$$(ii) \quad r_1 + r_3 = b_1 + 2b_2 + a_1 - a_2 \,.$$

Consider point $A(a_1, a_2, a_3)$; we know that $a_1 + a_3 \equiv 0 \pmod 3$, and that $a_3 = a_1 + a_2$. Hence $2a_1 + a_2 \equiv 0 \pmod 3$. Adding and subtracting a_1 we get:
$$3a_1 + (a_2 - a_1) \equiv 0 \rightarrow (a_2 - a_1) \equiv 0 \equiv (a_1 - a_2) \pmod 3 \,.$$

Also, since $(2a_1 + a_2) + (a_1 + 2a_2) = 3(a_1 + a_2) \equiv 0 \pmod 3$ we deduce that $a_1 + 2a_2 \equiv 0 \pmod 3$.

The same deductions can be made about $B(b_1, b_2, b_3)$, and indeed any B-point. Hence $3|[(a_1 + 2a_2) + (b_1 - b_2)]$, and $3|[(b_1 + 2b_2) + (a_1 - a_2)]$, so both L and R are B-points. $\square$

It is evident that if ABC is an integer equilateral triangle, each vertex. is an ET-transform of the other pair. We use this fact in the next theorem.

Theorem 5.4: The centroid of an equilateral triangle of B-points in π_0 is a lattice point.

Proof: Let A, B, C be the B-point vertices of the equilateral triangle. Suppose that $A \rightarrow B \rightarrow C$ is anticlockwise. Then:

$$C = E(AB) = (a_1 + a_2 - b_2, b_1 + b_2 - a_1, a_2 + b_1)$$

Hence, if G is the centroid of $\triangle ABC$:

$$\begin{aligned} 3G &= (a_1 + b_1 + a_1 + a_2 - b_2, a_2 + b_2 + b_1 + b_2 - a_1, \\ &\qquad a_1 + 2a_2 + 2b_1 + b_2) \\ &= [(2a_1 + a_2) + (b_1 - b_2), (a_1 - a_2) + (b_1 + 2b_2), \\ &\qquad (a_1 + 2a_2) + (2b_1 + b_2)] . \end{aligned}$$

We showed in the proof of Theorem 5.3 that all the expressions in round brackets, in the above coordinates, are congruent to 0 (mod 3). It follows that, dividing through by 3, all the coordinates of G are integer, and hence G is a lattice point. $\square$

5.3   Study of Triangles in π_0, using ET-transforms

We have seen how certain triangles with integer vertices can be drawn in the honeycomb plane, whilst others cannot. We can use the ET-transform and the theorems obtained above, to discover other triangles in these categories. We can also apply the ET-transform to the sides of triangles (or other polygons), and study the geometric figures that can be drawn on the resulting points. Many interesting questions can be asked about these figures.

5.31 Some triangle constructions

The diagrams below illustrate the constructions which are now to be described.

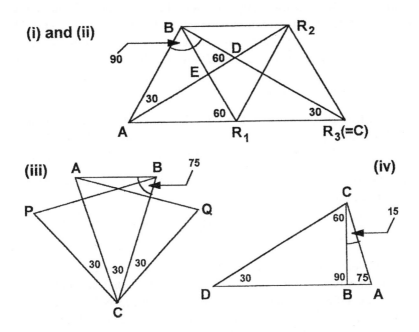

Figure 1. Some triangle constructions in plane x+y=z

(i) Given a segment AB. Construct a 60° **rhombus** on AB, and an integer **(90,60,30) right-triangle** with AB the short arm.

Method: Determine points R_1, R_2, $R_3(=C)$ by three consecutive ET-transforms: $R_1 = E(BA)$, $R_2 = E(BR_1)$, $R_3 = C = E(R_2R_1)$.

$$
\begin{aligned}
R_1 &= (b_1 + b_2 - a_2,\ a_1 + a_2 - b_1,\ b_2 + a_1) \\
R_2 &= (2b_1 + b_2 - a_1 - a_2,\ a_1 + b_2 - b_1,\ b_1 + 2b_2 - a_2) \\
R_3 &= (2b_1 + 2b_2 - a_1 - 2a_2,\ 2a_1 + a_2 - 2b_1,\ -3a_1 - a_2 + 2b_2) .
\end{aligned}
$$

Then ABR_2R_1 is the required rhombus; and ABR_3A is the required triangle. It is evident that in each case there are several other solutions.

(ii) Construct an integer **(90, 60, 30) right-triangle** with the given segment AB being the long arm. And another with AB the hypotenuse (state conditions).

Method: We can use figure (i) again. Let AR_2 and BR_3 intersect in D. Then $\triangle ABD$ is the required triangle if D (which is the centroid of $\triangle BR_1R_2$) is an integer point. From theorems 5.3 and 5.4, we know that D is an integer point if both A and B are B-points. In fact, D will be an integer point whenever (iff) $(a_2 - a_1) \equiv (b_2 - b_1)$ modulo 3. Examples of the three possible cases are:

(1) $A(1,4,5), B(2,5,7)$ $\rightarrow$ $D = (3,4,7)$ (A, B, R_1 are B-points)
(2) $A(1,2,3), B(2,3,5)$ $\rightarrow$ $D = (3,2,5)$ $(a_2 - a_1 \equiv +1, b_2 - b_1 \equiv +1)$
(3) $A(1,3,4), B(2,4,6)$ $\rightarrow$ $D = (3,3,6)$ $(a_2 - a_1 \equiv -1, b_2 - b_1 \equiv -1)$

It is easy to examine all possible cases modulo 3, to prove the above statement, using the fact that:

$$3D = [(-a_1 - 2a_2 + 4b_1 + 2b_2), (2a_1 + a_2 - 2b_1 + 2b_2),$$
$$(a_1 - a_2 + 2b_1 + 4b_2)]$$

For example, to prove case (2) we set: $a_2 = a_1 + 1 + 3j$ and $b_2 = b_1 + 1 + 3k$ into the first two coordinates of $3D$, and find that they are both exactly divisible by 3, for any values of a_1, b_1, j, k. We can proceed similarly to test cases (1) and (3), and also those for which A and B differ in their respective values of $a_2 - a_1$ and $b_2 - b_1$ modulo 3.

To make an integer, $(90°, 60°, 30°)$ right-triangle having AB as hypotenuse, we see in figure (ii) that $\triangle ABE$ has the required properties if E is an integer point. Now, $E = (B + R_1)/2$. Using the formulae for B and R_1, we find that E is an integer point only when $a_1 \equiv b_1 \pmod 2$ and $a_2 \equiv b_2 \pmod 2$.

(iii) Show that a $(30°, 75°, 75°)$ integer triangle is impossible.

Demonstration: In figure (iii), P and Q are ET-transforms of BC and CA respectively of a $(30°, 75°, 75°)$ triangle ABC. We show, by a contradiction, that triangle ABC is not an integer triangle.
Suppose that it were. Then by theorem 5.2 both P and Q are integer points. Also C is an integer point (given). Now $\angle ACB = 30°$ and $\angle PCB = 60°$, which implies that $\angle PCA = 30°$. Similarly $\angle BCQ = 30°$ and therefore $\angle PCQ = 90°$. Finally, since $CP = CA = CB = CQ$, we have that $\triangle PCQ$

is a $(90°, 45°, 45°)$ integer triangle. But this is impossible, by Theorem 5.1. Hence $\triangle ABC$ is not an integer triangle.

(iv) Show that a $(90°, 75°, 15°)$ integer triangle is impossible in π_0.

Demonstration: Suppose one were possible, drawn as in figure (iv). Then using construction (i), we can construct a $(90°, 60°, 30°)$ integer-triangle BCD having BC as short side, as shown in the diagram.
But that gives a combined integer triangle $\triangle DAC$ which is $(30°, 75°, 75°)$. But we have already shown that such an integer triangle is impossible. Hence the triangle $\triangle ABC$ shown in figure (iv) cannot have all integer vertices.

5.32 The ET-transform set $\mathcal{E}$ of a triangle

Suppose we are given an integer triangle $\triangle ABC$. Then from each side we can find two ET-transform points. We shall call a transform point an *outer* one if the perpendicular from it to its associated side lies wholly outside $\triangle ABC$. The others will be called *inner* transform points. We designate the transform points as follows: From sides AB, BC, CA the outers are respectively P, Q, R, and the inners are respectively P', Q', R'. We shall call the following set of integer points
$$\mathcal{E} = \{A, B, C, P, Q, R, P', Q', R'\}$$ the *ET-transform set* of $\triangle ABC$.

Now we can pose various types of problem about the set $\mathcal{E}$, for given types of integer triangle $\triangle ABC$. And we shall find that in order to answer them, we can resort to a mixture of geometric and number theoretic methods.
For example, we can ask what kinds of triangle are PQR and P', Q', R', and compare them with $\triangle ABC$.
There is one special triangle for which the answer to that is immediate: if ABC is equilateral, then PQR is also equilateral (with double the side-length) whilst P', Q', R' is an equilateral triangle which coincides with $\triangle CAB$.

More generally, we can ask which triples of points in $\mathcal{E}$ are collinear (can more than three points be collinear?); and which triples form triangles which are similar to $\triangle ABC$. The following two examples deal with these questions, for special cases of triangles.

The *ET*-transform set of (90,60,30) triangles

The diagram shows $\triangle ABC$, a general $(90°, 60°, 30°)$, together with all its six *ET*-transform points. From the diagram we glean the following information:

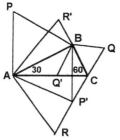

RQ'R' and *PQ'P'* are collinear triples, their lines intersecting at Q', the mid-point of AC. The following eight triangles are congruent to *ABC*:

$$AQ'R,\ CQ'R, P'BQ, AP'R,$$
$$AP'C, APR', AQ'R', CQ'R'.$$

Apart from the equilateral triangles used in the *ET*-transforms, $\triangle R'QQ'$ is also equilateral, with B as its centroid.

The sets of four points A, P, R', B and A, B, C, P' each forms a cyclic quadrilateral.

All of these lines and figures can be specified generally, using the coordinates from $A(a_1, a_2, a_3)$ and $B(b_1, b_2, b_3)$.

Constructing triangles *ABC*, for which P', Q', R' are collinear

It is possible for the three inner points in $\mathcal{E}$ to be collinear, as the following right-hand diagram shows. Note that the triangle ABC is isosceles. We shall show how to find the correct point C, when any two points A, B are given. And we conjecture that only an isosceles triangle similar to that constructed can have this property.

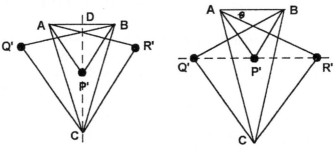

Figure 2. Diagrams for Theorem 5.5

Theorem 5.5: Given any two integer points, say $A(a_1, a_2, a_3)$ and $B(b_1, b_2, b_3)$. Then a point C can be found on the perpendicular bisector of AB such that the three inner transform points of $\triangle ABC$ are collinear.

Proof: Let C be any point on the perpendicular bisector of AB, as shown in the left-hand diagram. Let the ET-transforms of AC and CB be respectively R' and Q'. And let $\angle BAR' = \angle ABQ' = \theta$. The inner transform of AB is P', and this point must lie on DC, by symmetry.

If C be moved downwards from D, angle θ will increase; points Q' and R' will move apart, but the line joining them will remain parallel to AB. And since θ will increase, this line will move downwards. Hence, since P' remains fixed, there will be one value for θ at which $Q'R'$ will contain P'. Then the inner tranform points of $\triangle ABC$ will be collinear. $\square$

A formula for the point C

It is evident that a necessary and sufficient condition for the segment $Q'R'$ to pass through P' is that P' is the mid-point of $Q'R'$.

Using this fact, we compute formulae for the coordinates of C as follows. By the ET-transform formulae,

$$
\begin{aligned}
P' &= (b_1 + b_2 - a_2, a_1 + a_2 - b_1, b_2 + a_1) \\
Q' &= (c_1 + c_2 - b_2, b_1 + b_2 - c_1, c_2 + b_1) \\
R' &= (a_1 + a_2 - c_2, c_1 + c_2 - a_1, a_2 + c_1) \\
Q' + R' &= (a_1 + a_2 + c_1 - b_2, b_1 + b_2 - a_1 + c_2, a_2 + b_1 + c_1 + c_2)
\end{aligned}
$$

Applying the condition $Q' + R' = 2P'$, and equating the first two coordinates, we can solve for c_1 and c_2 to get the required coordinates of C:

$$
\begin{aligned}
c_1 &= 2b_1 + 3b_2 - a_1 - 3a_2 \\
c_2 &= 3a_1 + 2a_2 - 3b_1 - b_2 \\
c_3 &= c_1 + c_2 \,.
\end{aligned}
$$

There is another point C which satisfies our initial requirement, which of course is the mirror image (in AB) of the above solution.

Notes:

(i) We conjecture that the above solutions, obtained from isosceles triangles with coordinates calculated by the formulae, are the only possible solutions to the problem of finding a triangle whose inner ET-transform points are collinear.

(ii) Adding the coordinates of A, B, C for the solution triangle, we find that each coordinate is divisible by 3, and the resulting point is P'. Thus we have found that P' is the centroid of the solution triangle.

(iii) The sides of the solution triangle are in proportions $1 : \sqrt{7} : \sqrt{7}$. This is found by computing the lengths of the vectors $A - B$ and $B - C$, algebraically from the coordinates of the formulae for A, B and C, thus:

$$
\begin{aligned}
BC^2 &= |\mathbf{B} - \mathbf{C}|^2 \\
&= (2a_1 + 3a_2 - 2b_1 - 3b_2)^2 \\
&\quad + (3a_1 + a_2 - 3b_1 - b_2)^2 + (a_1 - 2a_2 - b_1 + 2b_2)^2 \\
&= a_1^2(4 + 9 + 1) + a_1 a_2(12 + 6 - 4) + a_2^2(9 + 1 + 4) \\
&\quad + b_1^2(4 + 9 + 1) + b_1 b_2(12 + 6 - 4) + b_2^2(9 + 1 + 4) \\
&\quad - a_1 b_1(8 + 18 + 2) - a_1 b_2(12 + 6 - 4) \\
&\quad - a_2 b_1(12 + 6 - 4) - a_2 b_2(18 = 2 + 8) \\
&= 7 \times 2(a_1^2 + a_1 a_2 + a_2^2 + b_1^2 + b_1 b_2 + b_2^2 \\
&\qquad\qquad\qquad -2a_1 b_1 - a_1 b_2 - a_2 b_1 - 2a_2 b_2) \\
&= 7|\mathbf{A} - \mathbf{B}|^2 = 7AB^2 , \quad \text{which was to be demonstrated.}
\end{aligned}
$$

5.33 An associated Diophantine equation

We can link the triangle calculations from the above subsection with number theory in an immediate, and satisfying, way, as follows.

The equation just used to find the squares of sides of $(1, \sqrt{7}, \sqrt{7})$ triangles suggests the Diophantine problem: Find all solutions in natural numbers of the equation

$$
x^2 + y^2 + z^2 = 7(u^2 + v^2 + w^2) \tag{*}
$$

The parametric solution (possibly not for *all* solutions, but we believe it might be so) is: Take any four natural numbers (p, q, r, s) and compute (x, y, z, u, v, w) by the following formulae (obtained from above, by setting $(a_1, a_2, b_1, b_2) = (p, q, r, s)$):

$$
\begin{aligned}
x &= (2p + 3q) - (2r + 3s) \\
y &= (3p + q) - (3r + s) \\
z &= (p - 2q) - (r - 2s)
\end{aligned}
$$

$$u = (p - r)$$
$$v = (q - s)$$
$$w = (p + q) - (r + s).$$

We can express this transformation in matrix form thus:

$$(x, y, z, u, v, w) = \begin{bmatrix} 1 & 3 & -1 & -3 \\ -3 & -2 & 3 & 2 \\ -2 & 1 & 2 & -1 \\ 1 & 0 & -1 & 0 \\ 0 & 1 & 0 & -1 \\ 1 & 1 & -1 & -1 \end{bmatrix} \begin{pmatrix} p \\ q \\ r \\ s \end{pmatrix}$$

Examples and notes

(i) Six solutions:

$$(4, 1, 5, 1, 2, 1), \quad (7, 7, 0, 2, 3, 1), \quad (7, 14, 7, 4, 5, 1)$$
$$(3, 1, 2, 0, 1, 1), \quad (10, 8, 2, 4, 2, 2), \quad (18, 13, 5, 3, 4, 7)$$

(ii) For any given solution, there are 35 others which are equivalent up to permutations of the x, y, z terms and the u, v, w terms.

For example, $(4, 1, 5, 1, 2, 1) \equiv (1, 4, 5, 1, 1, 2)$.

(iii) We define a primitive solution in the usual way. A solution is *primitive* iff $\gcd(x, y, z, u, v, w) = 1$. In examples (i), all solutions are primitive except $(10, 8, 2, 4, 2, 2)$; this is permutation-equivalent to $2(4, 1, 5, 1, 2, 1)$.

(iv) It is curious, with a Diophantine equation of this type, that sometimes two solutions can be added to form another one. For example: $(4, 1, 5, 1, 1, 2) + (1, 2, 3, 0, 1, 1) = (5, 3, 8, 1, 2, 3)$ is a solution to equation $(*)$. But $(4, 1, 5, 1, 2, 1) + (3, 1, 2, 0, 1, 1) = (7, 2, 7, 1, 3, 2)$ is not a solution.

A glance at the sum solution in (iv), viz. $(5, 3, 8, 1, 2, 3)$, shows that its terms are all Fibonacci numbers. It is natural, since we are working in the honeycomb plane, that we ask if there are classes of solutions of $(*)$ which either are Fibonacci vectors (6-tuples), or which can be characterised in terms of the Fibonacci numbers. One such infinite class of solutions of $(*)$ is the following, $\forall n \in \mathbf{Z}$:

$$\{(x, y, z, u, v, w) = (F_{n+3} + F_{n+1}, F_{n+4} - F_{n-1}, F_{n-2}, F_n, F_{n+1}, F_{n+2})\}$$

. This is obtained by putting $(p, q, r, s) = (F_{n+2}, F_{n+3}, F_{n+1}, F_{n+2})$ in the parametric form of the solution.

Expressing the result in another way, we can say that

$$S_n \equiv (F_{n+3} + F_{n+1})^2 + (F_{n+4} - F_{n+1})^2 + F_{n-2}^2$$

is always divisible by 7, and also by

$$T_n \equiv (F_n^2 + F_{n+1}^2 + F_{n+2}^2): \quad \text{and} \quad S_n = 7T_n.$$

This version says something interesting about sums of squares of three consecutive Fibonacci numbers.

Further, 14 divides S_n, since T_n is even (because two of any three consecutive Fibonacci numbers are odd, and the other is even).

One observation is that $S_1 = 2.7^2$ and $S_4 = 2.7^3$; we ask how many S_n are of the form 2.7^i.

5.34 *ET*-transforms of Fibonacci vector polygons

Consider the basic Fibonacci vector polygon in π_0, drawn on the vertices

$$\cdots, \mathbf{F}_1(0, 1, 1), \mathbf{F}_2(1, 1, 2), \mathbf{F}_3(1, 2, 3), \mathbf{F}_4(2, 3, 5), \mathbf{F}_5(3, 5, 8), \cdots$$

Taking the left *ET*-transforms of the polygon's sides, we get:

$$\begin{aligned}
E_L(\mathbf{F}_1\mathbf{F}_2) &= (0, 2, 2) \\
E_L(\mathbf{F}_2\mathbf{F}_3) &= (0, 2, 2) \\
E_L(\mathbf{F}_3\mathbf{F}_4) &= (0, 4, 4) \\
E_L(\mathbf{F}_4\mathbf{F}_5) &= (0, 6, 6)
\end{aligned}$$

It quickly becomes apparent, when a few more sides are *ET*-transformed, that all of the sides of the Fibonacci vector polygon are transformed onto the Y-axis; and that the points are themselves in a Fibonacci vector sequence generated by $\mathbf{F}(\mathbf{a}, \mathbf{b})$ with $\mathbf{a} = (0, 2, 2)$ and $\mathbf{b} = (0, 2, 2)$.

The corresponding right *ET*-transforms are:

$$\cdots, (1, 0, 1), (2, 1, 3), (3, 1, 4), (5, 2, 7), \cdots$$

We note that this is another Fibonacci vector sequence, with initial vectors $(1, 0, 1)$ and $(2, 1, 3)$; but it is NOT a vector sequence inherent to plane π_0.

If we take the left *ET*-transform of this new vector sequence, we find a small surprise. It does not return us to the original vector sequence, but to one which is *twice the original*. Symbolically, $E_L(E_R(\mathcal{F})) = 2\mathcal{F}$.

We shall not pursue these studies of ET-transforms of Fibonacci vector polygons any further, although it is evident that many interesting results await discovery. We conclude the Chapter by stating and proving the general result for left ET-transforms of inherent Fibonacci vector sequences, as exemplified above.

Theorem 5.6: Let $\mathbf{F}(\mathbf{a}, \mathbf{b})$ be an inherent Fibonacci vector sequence in π_0. Then the left ET-transform of its vector polygon is a linear Fibonacci vector sequence in the Y-axis.

Proof: Let $\mathbf{a} = (a_1, a_2, a_1 + a_2)$. Then, for an inherent vector sequence in π_0, we have $\mathbf{b} = (a_2, a_1 + a_2, a_1 + 2a_2)$. And the next two terms are:

$$\begin{aligned} \mathbf{c} &= (a_1 + a_2, a_1 + 2a_2, 2a_1 + 3a_2) \\ \mathbf{d} &= (a_1 + 2a_2, 2a_1 + 3a_2, 3a_1 + 5a_2) . \end{aligned}$$

Taking left ET-transforms, we obtain:

$$\begin{aligned} E_L(\mathbf{ab}) &= (0, 2a_2, 2a_2) &= 2a_2(0, 1, 1) \\ E_L(\mathbf{bc}) &= (0, 2(a_1 + a_2), 2(a_1 + a_2)) &= 2(a_1 + a_2)(0, 1, 1) \\ E_L(\mathbf{ca}) &= (0, 2(a_1 + 2a_2), 2(a_1 + 2a_2)) &= 2(a_1 + 2a_2)(0, 1, 1) \end{aligned}$$

It is evident that this is a sequence on the Y-axis, with the general term being $2(F_n a_1 + F_{n+1} a_2)(0, 1, 1)$. We should show this to be true by taking $\mathbf{c}$, $\mathbf{d}$ as the nth and $(n+1)$st terms respectively of $\mathbf{F}(\mathbf{ab})$, and taking the ET-transform of $\mathbf{cd}$. It is an elementary exercise to do this. $\qquad\square$

As a final comment, we remark that it is curious that we have found a way to transform all of the inherent Fibonacci vector polygons onto the same straight line (the Y-axis), when the start of all our investigations in this Part was to 'lift' all of the Fibonacci number sequences (by 'vectorizing') out of the number line and spread them out into the plane π_0.

Chapter 6

Vector Sequences Generated in Planes

In Chapters 2 and 3 we defined integer-vector recurrence equations, and studied some of the geometric properties of vector sequences generated by them. Figure 2 of Chapter 3 shows the honeycomb plane $\pi_0 \equiv z = x + y$ with parts of the basic Fibonacci and Lucas vector polygons plotted on it.

In this Chapter we shall develop these ideas further. In particular, we shall study vector sequences in the general plane $z = cx + dy$, defining first a vector/matrix mode of generating vector sequences which are 'inherent to the plane'. We shall call the equation for this generating mode the general Type I vector recurrence equation. And the vector recurrence relation defined in Chapter 2 will now be called a Type II recurrence equation. We begin by listing eight examples of integer vector sequences, and discussing properties of them. Then we state and prove several general theorems about Type I and Type II generated vector sequences.

Finally, we concentrate upon the class of Fibonacci vector sequences in the honeycomb plane, and study various aspects and transformations of them.

6.1 Eight Example Vector Sequences

We recall the following terminology: $\mathbf{x} = (x, y, z)$ is an *integer vector* if all of its coordinates are integers; we shall say it is a *coprime vector* if also $\gcd(x, y, z) = 1$. We shall generally omit the adjective 'integer' when it is clear that integer vectors are being treated.

$\{\mathbf{x}_n\}$ will denote an *integer vector sequence* if $\mathbf{x}_n$ is an integer vector for each given value of n. It will be a coprime vector sequence if all its terms are coprime vectors.

Eight examples of integer vector sequences follow, with discussion on some of their properties.

Examples of vector sequences:

(1) Basic Fibonacci:
$$(0,1,1), \ (1,1,2), \ (1,2,3), \ ..., \ (F_{n-1}, F_n, F_{n+1}), \ ...$$
(2) Basic Lucas:
$$(2,1,3), \ (1,3,4), \ (3,4,7), \ ..., \ (L_{n-1}, L_n, L_{n+1}), \ ...$$
(3) Basic Pell:
$$(0,1,2), \ (1,2,5), \ (2,5,12), \ ..., \ (P_{n-1}, P_n, P_{n+1}), \ ...$$
(4) $(0,1,1), \ (1,2,3), \ (3,5,8), \ ..., \ (F_{2n-2}, F_{2n-1}, F_{2n-2}), \ ...$
(5) $(0,1,1), \ (2,3,5), \ (8,13,21), \ ..., \ (F_{3n-3}, F_{3n-2}, F_{3n-1}), \ ...$
(6) $(0,1,2), \ (1,1,3), \ (1,2,5), \ ..., \ (F_{n-1}, F_n, F_{n+2}), \ ...$
(7) $(0,1,3), \ (1,3,8), \ (3,8,21), \ (8,21,55), \ ..., \ (F_{2n-2}, F_{2n}, F_{2n+2}), \ ...$
(8) $(1,1,2), \ (3,5,8), \ (4,6,10), \ (7,11,18), \ (11,17,28), \ ...$

We observe that vector sequences (1), (2) and (3) are, respectively, the basic Fibonacci, Lucas and Pell sequences. (4) is a subsequence of (1), and in vector form it is: $\mathbf{F}_1, \mathbf{F}_3, \mathbf{F}_5, ..., \{\mathbf{F}_{2n-1}\}, ...$ All eight are integer vector sequences: only example (8) is **not** a coprime sequence.

It is of interest to note that in both the vector sequences (4) and (5), the union set of all coordinates used in the vectors is the set of Fibonacci numbers $\mathcal{F} = \{0, 1, 1, 2, 3, 5, 8, ...\}$. In (5) the set $\mathcal{F}$ is used exactly, each Fibonacci number appearing once only among the coordinates; whereas in (4) there is redundancy — the set $\{F_{2n}\}$ is used twice. A problem for study is to determine other vector sequences which use only Fibonacci numbers as coordinates; in which planes these sequences occur, and how much redundancy they have.

Vectors in (6) use only Fibonacci numbers, but they do not lie in the Fibonacci plane. Vector sequence (7) is another example of such sequences; this time the vectors all lie in the plane $z = -x + 3y$, and there is both deficiency and redundancy in the use of the Fibonacci numbers in their coordinates. Other examples of vector sequences with these properties will be given below.

Finally, vectors in sequence (8) lie in the same plane as do those of sequence (1); but only the first two vectors have all their coordinates Fibonacci numbers.

6.2 Vector Sequence Planes

The vectors in sequences (1) to (8) above lie in planes of type $z = cx + dy$. Thus vectors of (1) and (2) are all in the plane $\pi_0 \equiv z = 1x + 1y$. Vectors of (3) lie in the plane $z = 1x + 2y$; we shall refer to it as *the Pell plane*, and designate the nth vector in the sequence by $\mathbf{P}_n$.

We can use $\pi(c, d)$ to symbolize the planes $z = cx + dy$. Then the Fibonacci and Pell planes are, respectively, $\pi(1, 1)$ and $\pi(1, 2)$. The only other plane represented in the examples is $\pi(-1, 3)$, which includes sequence (7). We shall call the planes $\pi(c, d)$, when c and d are both integers, *sequence planes*; we shall see below how vector sequences can be generated in them in a special way, which determines their *inherent vector sequences*.

Before moving on to discuss vector recurrence relations, we include an example of an integer vector sequence which is **not** in a sequence plane as just defined, thus:

(9)   (1,2,2), (3,2,3), (4,4,5), (7,6,8), ..., $(L_n, 2F_n, F_{n+2})$, ...

The vectors in this sequence all lie in the plane $2x + 3y = 4z$, which is not a sequence plane since it cannot be expressed in the form $z = cx + dy$ with integer coefficients..

In Section 6.4 we shall define two types of recurrence relation, by which the vector sequences exemplified above are generated. First we must look at certain transformation matrices, which are associated with the plane equations in interesting ways.

6.3 Inherent Transformations of Planes

Suppose that the vector $\mathbf{x} = (x, y, z)$ belongs to the plane $\pi(c, d)$. Then we call* the 3×3 matrix H the *inherent transformation matrix* of the plane if $\mathbf{x}' = (y, z, cy + dz)$ and if $\mathbf{x}H = \mathbf{x}'$; that is, if H transforms $\mathbf{x}$ into the given $\mathbf{x}'$. We call this a *forward transformation* of $\mathbf{x}$. It may quickly be checked that:

$$H = \begin{pmatrix} 0 & 0 & 0 \\ 1 & 0 & c \\ 0 & 1 & d \end{pmatrix}.$$

---

*We use the symbol H for this matrix as a tribute to the Australian mathematician A. F. Horadam, who has published so many beautiful results about pairs of Fibonacci sequences.

We note that $\det(H) = 0$, so H is a singular matrix; therefore it has no inverse. However, we can choose (see below) a pseudo-inverse H^- which will perform a *backward transformation* of the kind we want. The desired pseudo-inverse is one which satisfies $\mathbf{x}'H^- = \mathbf{x}$; it has the following form, as may be checked easily:

$$H^- = \begin{pmatrix} -d/c & 1 & 0 \\ 1/c & 0 & 1 \\ 0 & 0 & 0 \end{pmatrix}.$$

To see that H and H^- are pseudo-inverses of one another, it is only necessary to check that both have zero determinant and that $HH^-H = H$ and $H^-HH^- = H^-$. Thus, the matrix H^- is a reflexive generalized inverse of H.

The inherent matrix $H(1,1)$ for the honeycomb plane has the following properties, in its powers H^n and H^{-n}.

Theorem 6.1

$$(i) \quad H^n = \begin{pmatrix} 0 & 0 & 0 \\ F_{n-2} & F_{n-1} & F_n \\ F_{n-1} & F_n & F_{n+1} \end{pmatrix};$$

$$H^{-n} = \begin{pmatrix} F_{-n-1} & F_{-n} & F_{-n+1} \\ F_{-n} & F_{-n+1} & F_{-n+2} \\ 0 & 0 & 0 \end{pmatrix}.$$

$$(ii) \quad H^n + H^{-n} = \begin{pmatrix} \mathbf{F}_{-n} \\ \mathbf{F}_{1-n} + \mathbf{F}_{1+n} \\ \mathbf{F}_n \end{pmatrix}, \text{ for } n = 1, 2, 3, \ldots$$

[Recall that $\mathbf{F}_n = (F_{n-1}, F_n, F_{n+1})$.]

It is easy to show that the middle row of this matrix is $L_{n-1}\mathbf{F}_0$ if n is odd, and $F_{n-1}\mathbf{L}_0$ if n is even.

(iii) The characteristic polynomial of H^n is $-\lambda(\lambda - \alpha^n)(\lambda - \beta^n)$; hence its characteristic roots are 0, α^n, β^n where α is the golden ratio, and $\alpha\beta = -1$.

Proof (i): Both formulae, for H^n and H^{-n}, are proved easily using induction on n. $\qquad\square$

Proof (ii): Using (i), only trivial checking of the terms of the formulae is required. □

Proof (iii): Using H^n from (i), and I12, I13 from [12, p. 57] we have:

$$|H^n - \lambda I| = \begin{vmatrix} -\lambda & 0 & 0 \\ F_{n-2} & F_{n-1} - \lambda & F_n \\ F_{n-1} & F_n & F_{n+1} - \lambda \end{vmatrix}$$

$$= -\lambda(\lambda^2 - L_n\lambda + (-1)^n) .$$

The roots of the quadratic factor in the above expression are $\lambda = L_n \pm \sqrt{L_n^2 - 4(-1)^n})/2 = (L_n \pm F_n\sqrt{5})/2$, (using identity I12). Hence, using $\alpha^n = F_{n-1} + F_n\alpha$, we obtain the required results: the characteristic roots are $0, \alpha^n, \beta^n$. □

Now we are ready to define our two types of recurrence relation, for generating vector sequences.

6.4 Vector Recurrence Relations

6.41 Type I: The matrix/vector equation, order 1
We shall study how we can use inherent transition matrices (and others that are closely related to them) to generate the kinds of vector sequence which we gave as examples in Section 6.1.

Using H and H^-, for any given sequence plane, we can generate vector sequences which are entirely in that plane, if we take a starting vector which is also in the plane. The two recurrence equations which we need to do this are:

$$\begin{array}{llll} Forward: & \mathbf{x}_{n+1} & = & \mathbf{x}_n H(c,d) & (6.1) \\ Backward: & \mathbf{x}_{n-1} & = & \mathbf{x}_n H^-(c,d) & (6.2) \end{array}$$

These two relations, together with a starting vector $\mathbf{x}_1$ in $\pi(c,d)$, will generate a doubly-infinite vector sequence, all in that same plane.

Looking back to the examples (1) to (8) in Section 6.1, we see the following illustrations of (6.1) and (6.2) in operation. Thus:

(1)′ Using $H(1,1)$, with $\mathbf{x}_1 = (0,1,1)$, we get the forward sequence shown. And, using $H^-(1,1)$, the following backward sequence is produced:

$\leftarrow \ldots, (-2,1,-1), (1,-1,0), (-1,0,1), (0,1,1)$. The two together comprise the full, basic, Fibonacci vector sequence.

(2)′ Similarly, using $H(1,1)$ and $H^-(1,1)$, with $\mathbf{x}_1 = (2,1,3)$, the two-way basic Lucas vector sequence is generated.

(3)′ The two-way Pell vector sequence shown is generated using $H(1,2)$ and its inverse, together with $\mathbf{x}_1 = (1,1,3)$, in relations (6.1) and (6.2). In order to generate the *basic* Pell vector sequence, we must use initial vector $\mathbf{P}_n = (0,1,2)$ in the relations. We then obtain the vector sequence:

$\leftarrow \ldots (-2,1,0), (1,0,1), (0,1,2), (1,2,5), (2,5,12), \ldots \rightarrow$

(4)′ The vectors in the sequence (4) are the basic Fibonacci vectors taken two-apart. We could write this sequence thus:

$\qquad \mathbf{F}_1, \mathbf{F}_3, \mathbf{F}_5, \mathbf{F}_{2n-1}, \ldots \rightarrow$

The square of $H(1,1)$ will effect the transformations needed for this sequence. Similarly, $H^{-2}(1,1)$ will effect the backward transformations. Hence this two-way vector sequence is given by the following recurrence relations, of order 1:

$\qquad \mathbf{x}_{n+1} = \mathbf{x}_n H^2(1,1)$ and $\mathbf{x}_{n-1} = \mathbf{x}_n H^{-2}(1,1)$, again with $\mathbf{x}_1 = (0,1,1)$.

If we were to use $(1,1,2)$ as initial vector, we should generate the vector sequence of all even-subscripted basic Fibonacci vectors. Hence the union of these two subsequences would use up all the Fibonacci numbers, in their coordinates, three times.

We remark, looking back at Theorem 6.1, that we can generate the Fibonacci vectors n-apart by using $H^n(1,1)$ and its inverse; it seems most appropriate that their characteristic roots should include α^n and α^{-n}, linked to forward and backward generation respectively. However, it is not so surprising, for the general solution to the vector/matrix equation is intimately linked with the characteristic roots of H. The next theorem shows how.

Theorem 6.2: (General solution of inherent Type I, Binet form)

Let $\mathbf{x}_{n+1)} \in \pi(1,1)$ and $H = H(1,1)$. Then:

(i) $\quad \mathbf{x}_{n+1} \quad = \quad \frac{1}{\sqrt{5}}[\alpha^{n-2}H(I + \alpha H) - \beta^{n-2}H(I + \beta H)]\mathbf{x}_1$.

(ii) $\quad \mathbf{x}_{n+1} \quad \rightarrow \quad \frac{1}{\sqrt{5}}\alpha^{n-2}H(I + \alpha H)\mathbf{x}_1$, as $n \rightarrow \infty$.

Proof:

If we repeatedly apply the recurrence $\mathbf{x}_{n+1} = H\mathbf{x}_n$, for $n = 1, 2, 3, \ldots, n$, we obtain $\mathbf{x}_{n+1} = H^n\mathbf{x}_1$.

Inserting the Binet identity $F_n = (\alpha^n - \beta^n)/\sqrt{5}$ into the expression for H^n given in Theorem 6.1, we get the following:

$$\sqrt{5}H^n = \begin{pmatrix} 0 & 0 & 0 \\ \alpha^{n-2} & \alpha^{n-1} & \alpha^n \\ \alpha^{n-1} & \alpha^n & \alpha^{n+1} \end{pmatrix}$$
$$- \begin{pmatrix} 0 & 0 & 0 \\ \beta^{n-2} & \beta^{n-1} & \beta^n \\ \beta^{n-1} & \beta^n & \beta^{n+1} \end{pmatrix}$$

$$= \alpha^{n-2}H(I + \alpha H) - \beta^{n-2}H(I + \beta H) .$$

Part (i) of the theorem follows immediately. And, letting $n \rightarrow \infty$ and noting that $\beta < 1$, part (ii) also follows. $\square$

6.42 General Type I vector recurrences

The inherent transmission matrix H is a specially chosen matrix, related directly to the equation of the plane of operation. We can, however, replace H by any 3×3 integer matrix (say T), and choose the elements of T so that different kinds of vector sequence result from the recurrence relation. It is clear that setting rules for the choice of Ts elements is tantamount to defining classes of integer vector sequences. We shall call the recurrence equation: $\mathbf{x}_{n+1} = \mathbf{x}_n T$ the *general Type I vector recurrence*, and give one new example. [N.B.—the use of H^2 for T is an example already discussed.]

Example:

Working in the Pell plane π_0, with starting vector $\mathbf{x}_1 = (1, 2, 5)$, and transmission matrix:

$$T = \begin{pmatrix} 0 & 1 & 2 \\ 1 & 1 & 3 \\ 0 & 0 & 0 \end{pmatrix} \quad \text{with inverse } T^- = \begin{pmatrix} 0 & 0 & 0 \\ 3 & -2 & -1 \\ -1 & 1 & 1 \end{pmatrix},$$

the following double vector sequence is generated:

$$\leftarrow \ldots (1, 1, 3), (1, 2, 5), (2, 3, 8), (3, 5, 13), \ldots \rightarrow$$

This sequence uses the Fibonacci numbers twice, for its coordinates, with all the vectors being in the Pell plane. Some pleasing comparisons of T with $H(1, 1)$ can be made: for example, they have the same characteristic equation.

6.43 Type II: The Fibonacci vector recurrence equation, order 2

The Type II vector recurrence relation (in plane π_0) is simply the analogue of the one which generates the well-known Fibonacci numbers: the numbers in that relation become vectors, and the number addition operation becomes vector addition. Thus the Type II vector recurrence relative to plane π_0 is (see Def. 2.1, in Chap.2): $\mathbf{x}_{n+2} = \mathbf{x}_n + \mathbf{x}_{n+1}$. Although this is the vector recurrence equation relative to the Fibonacci plane, the vector sequence which it generates is always entirely in the plane which is determined by Q(0,0,0) and the two starting vectors.

Similarly, the Type II recurrence equation relative to plane $\pi(c, d)$ is:

$$\mathbf{x}_{n+2} = c\mathbf{x}_n + d\mathbf{x}_{n+1}.$$

For examples, we observe that each of the vector sequences (1), (2), (3), (5), and (7) may be generated by a Type II recurrence. Notice that, for example, vector sequence (3) is in the Pell plane, whereas it is generated by the Type II recurrence which is relative to the Fibonacci plane.

This type of vector sequence has been treated in earlier Chapters; for example, some geometric properties of it were given in Chapter 3. For now, we will give one useful theorem which relates Type I and Type II generations of vector sequences in the general plane $\pi(c, d)$.

Theorem 6.3: If $\mathbf{x} = (x, y, z)$ is in the plane $\pi(c, d)$, then the sequence generated by the Type I inherent recurrence $\mathbf{x}_{n+1} = \mathbf{x}_n H(c, d)$ is the same as that generated by the Type II relative recurrence of that plane, viz. $\mathbf{x}_{n+2} = c\mathbf{x}_n + d\mathbf{x}_{n+1}$.

Proof: The first three terms of the sequence generated by the Type I recurrence are $\mathbf{x}$, $\mathbf{x}H$, and $\mathbf{x}H^2$. Using the general form of the inherent matrix $H(c, d)$ (see Section 6.3), we find that $\mathbf{x}H^2 = c\mathbf{x} + d(\mathbf{x}H)$ if (and only if) $z = cx + dy$ — that is, if $\mathbf{x}$ is in the plane $\pi(c, d)$. $\qquad\square$

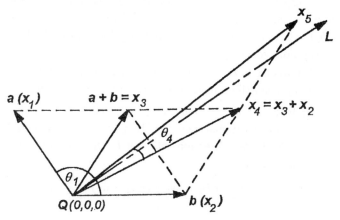

Figure 1. Type II production of vector sequence $\mathbf{G}(\mathbf{a}, \mathbf{b})$

6.5 Some Geometric Observations (continuation from Chapter 2)

The first subsection (6.51) begins with a recapitulation of the essential results presented in Chapter 2. It seems appropriate to give this before presenting further geometric findings on the whole class of Fibonacci vector polygons in the honeycomb plane, in 6.52 et seq. It would be wise, perhaps, to re-read Chapter 2 before continuing.

6.51 Demonstration 1 (recapitulation):

Figure 1 above shows the general development of a general Type II vector sequence, say $\mathbf{G}(\mathbf{a}, \mathbf{b})$. We assume that $\mathbf{a}$ and $\mathbf{b}$ are non-collinear, integer vectors directed from origin Q.

Observations (i):

Because of the repeated operation of the vector addition law, the angle θ_n reduces each time that n increases; it is clear that $\lim_{n\to\infty} \theta_n = 0$. Further, successive vectors in the sequence oscillate above and below a line through $Q(0,0,0)$. We call this the limit-line L. Note that all vectors with odd subscripts are above, and all with even subscripts are below, the line L. We can write, formally, that $\lim_{n\to\infty} \mathbf{x}_n = L$.

It is easy to show (see section 2.4), algebraically, that if $\mathbf{a}$ and $\mathbf{b}$ are any two vectors in $\mathbf{Z}^3$, then L has equations $x/(a_1 + \alpha b_1) = y/(a_2 + \alpha b_2) = z/(a_3 + \alpha b_3)$, where α is the golden ratio. If the sequence is in the honeycomb plane, then the following theorem holds.

Theorem 6.4: If $\mathbf{a}$ and $\mathbf{b}$ are consecutive Fibonacci vectors in π_0, then the limit-line L is independent of them.

Proof: Let $\mathbf{a} = (a_1, a_2, a_1 + a_2)$; then, since $\mathbf{a}$ and $\mathbf{b}$ are consecutive Fibonacci vectors, $\mathbf{b} = (a_2, a_1 + a_2, a_1 + 2a_2)$. Now, considering the direction-ratios of L, the ratio of the first two components is:

$$\frac{a_2 + \alpha b_2}{a_1 + \alpha b_1} = \frac{a_2 + \alpha(a_1 + a_2)}{a_1 + \alpha b_1} =$$

$$\frac{\alpha(a_1 + \alpha a_2)}{a_1 + \alpha a_2} = \alpha \ (\text{using } \alpha^2 = 1 + \alpha).$$

Similarly, it follows easily that the ratio of the second two components is also α. Hence direction-ratios of L are $1, \alpha, \alpha^2$, and so L is $x/1 = y/\alpha = z/\alpha^2$, independent of the two starting vectors. $\square$

Comment: This result is analogous to the fact that in any Fibonacci number sequence, the ratio F_{n+1}/F_n tends to α, independently of the choice of starting numbers.

Observations (ii):

(a) The sequence of triangles $Q\mathbf{x}_1, \mathbf{x}_2, \ Q\mathbf{x}_2, \mathbf{x}_3, \ Q\mathbf{x}_3, \mathbf{x}_4, \ \ldots$ have equal areas.

(b) The sequence of quadrilaterals $Q\mathbf{x}_1\mathbf{x}_3\mathbf{x}_2, \ Q\mathbf{x}_2\mathbf{x}_4\mathbf{x}_3, \ Q\mathbf{x}_3\mathbf{x}_5\mathbf{x}_4, \ \ldots$ are parallelograms of equal area.

(c) $x_1x_3x_4$, $x_2x_4x_5$, $x_3x_5x_6$, ... are straight-line segments, with x_3, x_4, x_5 etc. being their mid-points.

(d) All terms of the vector sequence are in the plane determined by the three points Q, a, b.

Observations (iii):

The well-known Simson's identities (see I13, [12, p. 57]) may be deduced directly from (ii)(a) and the geometry of the plane, thus:
—Suppose that $a = x_1$ and $b = x_2$ are consecutive Fibonacci vectors, in the plane $\pi_0 \equiv x + y - z = 0$. Then x_n and x_{n+1} lie in π_0, and the areas of all triangles Qx_nx_{n+1} are proportional to the magnitudes of the vector cross-products $x_n \times x_{n+1}$ (recall that $\Delta = \frac{1}{2}|x_n|.|x_{n+1}|\sin\theta$); but the cross-product coordinates are proportional to $(1, 1, -1)$, the direction-ratios of a normal to the plane π_0. Putting these two results together, we immediately get the Fibonacci identities:

$$F_{n-1}F_{n+1} - F_n^2 = (-1)^n = F_{n-1}F_{n+2} - F_nF_{n+1}$$

It is curious how the inductive process of arriving at proof of these identities is all taken care of by the geometry of the plane and the Type II vector sequence generated in it.

It is also of note that we can immediately generalize Simson's identities to other sequences, in other planes. For example, if a and b are both in the Pell plane $\pi(1, 2)$, we know that the normal coordinates are proportional to $(1, 2, -1)$ (the plane is $x + 2y - z = 0$), and there follow from this two identities in the basic Pell sequence of numbers.

6.52 Demonstration 2:

In Figure 2, Chapter 3, we showed how the Fibonacci and Lucas vector polygons zig-zagged upwards in the honeycomb plane, tending towards the limit line $L \equiv x/1 = y/\alpha = z\alpha^2$.

We now wish the reader to form in his/her mindspace (see [22]) a picture of the complete set of Fibonacci vector polygons in the plane π_0. Imagine that every integer vector in the plane $x + y = z$ simultaneously generates (by the inherent matrix/vector Type I recurrence equation) a Fibonacci vector sequence. Joining consecutive points by line-segments, we now have a 'mind-picture' of the complete set of Fibonacci vector polygons.

It should be clear that (i) each integer point lies on only one such polygon, and (ii) all the polygons will criss-cross each other infinitely often, but only at non-integer points; and (iii) we know that as they all progress 'upwards' they will tend to each other and the line L. The total result is a very complex networking of intersecting polygons in π_0 — a seemingly incomprehensible mish-mash of them.

To reduce the mental strain, instead of thinking of the formation of Fibonacci polygons, we can imagine a point map of the whole plane to itself. That is, imagine every point, such as $\mathbf{x}_n$, moving to its image point $\mathbf{x}_{n+1}$ under the Type I inherent transformation.

To help see where all the points go, we show in (a) below how the transformation $\mathbf{x}_{n+1} = \mathbf{x}_n H$, when applied simultaneously to all points of π_0, causes the plane to transform. The points move as a sequence of sectors, say $\{S_i\}$. Thus sector S_0 maps to sector S_1, which maps to sector S_2, and so on. The boundaries of these sectors are the lines (rays from $Q(0,0,0)$): $x/F_{n-1} = y/F_n = z/F_{n+1}$, with $n = 0, 1, 2, \ldots$. (These sectors are marked in Figure 2(a) below.)

Another helpful thing to do, is to think of the mapping obtained by using H^2 rather than H as the transformation matrix. With this, sectors map to alternate sectors thus: $S_1 \rightarrow S_3 \rightarrow S_5 \ldots$, and $S_0 \rightarrow S_2 \rightarrow S_4 \ldots$

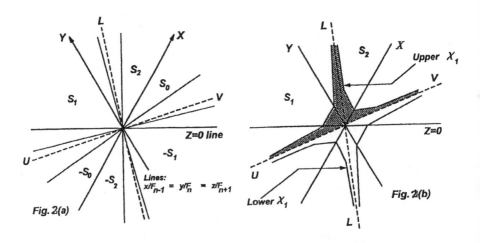

Figure 2(a). The points transformation of π_0, using H
2(b). The upper and lower chimneys $\pm\chi_1$

If we 'join up' the consecutive pairs of dots in *this* H^2 mapping, we find that each vector polygon is now separated into two branches, *with no crossings at all*. In Figure 2(b) we show the two branches which together contain the basic Fibonacci vector sequence. They form a kind of tilted funnel, or chimney, up which the basic Fibonacci polygon zig-zags. We have also included in the diagram the *negative* of this chimney — i.e. the mappings of all points $-\mathbf{F}_n$ under H^2.

Therefore, we propose to call the pair of branches the *upward basic Fibonacci Chimney*, and its negative the *downward basic Fibonacci chimney*. Similarly, we can imagine the chimneys formed from the vector sequence $\{2\mathbf{F}_n\}$, and from $\{3\mathbf{F}_n\}$, and so on, applying the H^2 transformation to all of their points. The result is a sequence of nested chimneys, which we shall designate by $\{\chi_m\}$. [The notation seems most appropriate: using 'chi-m' for 'chimney'.]

In Figure 3 below, we give two diagrams which show how the nests of chimneys are arranged in the plane π_0.

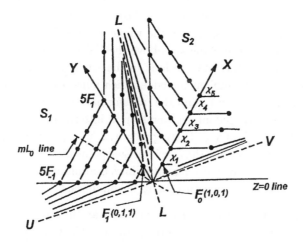

Figure 3(a). The first five upward Fibonacci chimneys

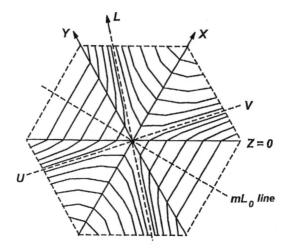

Figure 3(b). The Fibonacci chimney diagram (or web) (smoothed)

In Figure 3(a), we show the first four upper chimneys only. On the branches, we show points from other Fibonacci vector sequences; and we claim that in all of the branches of the whole set of chimneys, there lie all the integer-vectors in the plane π_0 (above the UV-limit-line).

To prove this, we only have to look in the sector S_1, and observe that there are m 'nearest points' on the edge which joins $m\mathbf{F}_{-1}$ to $m\mathbf{F}_1$ (in the left branch of that chimney). And the transformation by H takes this edge, and all of the evenly spaced integer points in it, into the image edge in sector S_2, in the right branch of χ_m. Letting $m = 1, 2, 3, \ldots$, and then generating all chimneys from S_1, we see that each integer point above UV will be sited on some Fibonacci chimney branch. It follows that all integer-vectors below UV will occur on the downward Fibonacci chimneys.

In Figure 3(b), we give a picture of the way in which the Fibonacci chimneys cover the integer-points of the plane π_0. The branches are actually piece-wise linear, but we have 'smoothed' their sides together, to produce a memorable diagram; the similarities to hyperbolic curves are striking. Note the asymmetry of the chimney directions with respect to the axes, and to limit-line UV. Note too how the branches of the chimneys are different on either side of limit-line L; but they each have symmetry axes, which are perpendicular.

This is our simplified, pictorial representation of the set of all Fibonacci vector sequences, to be stored in our mindspaces. (We have drawn in the first three central hexagons, dotted, to create a spider's web image of the system.)

A final remark is that we can think of the two branches of a chimney χ_m as being the envelope of all the m Fibonacci vector polygons which zig-zag upwards within the chimney.

6.53 Demonstration 3:

The Lucas vector polygon zig-zags upwards in the second Fibonacci chimney. Since $\mathbf{L}_0 = \frac{1}{2}(2\mathbf{F}_{-1} + 2\mathbf{F}_1)$ it is the mid-point of the first side (in sector S_1) of the left branch of the chimney.

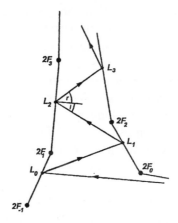

Figure 4. The chimney χ_2 and the Lucas polygon

Figure 4 shows the first four sides of the polygon. We have marked the incidence and reflection angles, respectively i and r, occurring when the polygon changes direction at point $\mathbf{L}_2$. To describe such changes of direction, when the vector polygon 'bounces off the chimney walls', we define the *chimney reflection (or zig-zag) ratio* to be: $f_n = b_n \sin r / (a_n \sin i)$. It can be computed for each reflection point, and the sequence $\{f_n\}$ provides a measure of the stretching and flattening-out of the polygon as n increases. Below the diagram, we give a formula for f_n, and note that f_n tends to α.

By taking the dot-products of the edges $\mathbf{L}_{n-1}\mathbf{L}_n$ and $\mathbf{L}_n\mathbf{L}_{n+1}$ with the chimney side $2\mathbf{F}_{n-1}2\mathbf{F}_{n+1}$, and forming the ratio of the results, we get the formula:

$$f_n = \frac{F_{2n+1} + F_{2n+3} + F_{2n+5} + (-1)^n}{F_{2n} + F_{2n+2} + F_{2n+4} + (-1)^{n-1}} \; .$$

It is easy to show[†] that $f_n \to \alpha$, the golden ratio, as $n \to \infty$ (which we could easily deduce by geometric observations on the rising Lucas vector polygon, but perhaps not rigorously).

6.6 Summary

In this Chapter we have defined two types of vector recurrence equation, and given many examples of vector sequences generated by them. These examples have all been related to Fibonacci sequences of one form or another; and we have tried to show that there is much of interest and value to be gained from studies of these. In particular, we have shown how inherent transmission matrices of a family of sequence planes generate vector sequences in those planes. We have given an attractive way of viewing (imagining) the whole set of Fibonacci vector sequences in the honeycomb plane $x + y - z = 0$; and we have demonstrated how the Lucas vector polygon 'rises' up its Fibonacci chimney.

We used the term 'transmission matrix' because we wish to encourage the notion of dynamic generation of vector sequences. Fibonacci vector sequences may be thought of as loci of points moving in zig-zag paths, up and down chimneys! This view has similarities with that usually taken in Markov Chain theory, where sequences of state probability vectors are generated using probability transition matrices. Perhaps fruitful parallels between Markov chain theory and Fibonacci vector geometry can be drawn.

[†]Divide numerator and denominator by F_{2n}, and let $n \to \infty$.

Chapter 7

Fibonacci Tracks, Groups and Plus-Minus Sequences

7.1 Introduction

During the past fifteen years J. Turner has had cause to study several types of sequences which were not of numbers but of other mathematical objects, such as trees [26], words [25], vectors [30], [31] and 2x2 matrices [28]. In each case they were generated by a generalized form of the Fibonacci recurrence equation: the objects concerned were defined in a set, and a binary relation for combining pairs of them was suitably defined; the symbol for this relation replaced + in the Fibonacci recurrence equation.

He began to believe that this procedure, and the resulting 'Fibonacci sequences', should have a generic definition and title; that the form itself was worthy of general study; and that, moreover, it could be a useful tool for studying the algebras of binary relations. In [29] he expressed these views, and proposed to use the name **track** for sequences which arose from the generalized form of Fibonacci recurrence. This name was intended to convey connotations suggested by some or all of the following words: 'sequence', 'path', 'process', 'route', 'stepping-stones', 'trajectory', 'trail', 'walk', 'sample', 'probe', 'trace', 'track in cloud chamber', and of course ' track' itself. Most of these words are well-used elsewhere in mathematics. In particular, the author thought that 'Fibonacci sequence' suggests 'Fibonacci number sequence' far too strongly for it to be retained for sequences generated from a form which is defined on any kind of object-set.

This chapter begins with a definition of the *Fibonacci track recurrence* form and follows with one example drawn from [29]. Then it develops the 'tool-use' of the form, by describing various studies of sets of tracks (called *tracksets*) which derive from binary algebras defined on finite sets, such as finite groups and latin-square algebras. The hope is expressed that at least some of the trackset properties discovered will shed new light on the binary algebras.

Finally, the methods of [31] are used to link studies of tracks in finite groups with vector geometry; and it is shown how a trackset of a group can be viewed as a knot.

7.2   Definition of a Track in a Set S

7.21 Definitions

General sequences of Fibonacci numbers are given by the linear second-order recurrence equation:

$$x_{n+2} = cx_{n+1} + dx_n \,, \text{ with } n = 0, 1, 2, \cdots$$

where c and d are fixed integers, and initial terms x_0 and x_1 are chosen from $\mathbf{Z}$.

In [29] the author proposed to remove the restriction that the terms x_n shall be integers (they may or may not be). All that is necessary is that they are elements of a given set $\mathbf{S}$, which has a binary operation $\oplus$ defined on $\mathbf{S} \times \mathbf{S}$. The coefficients c and d and 'scalar multiplication' must be suitably well-defined such that the linear forms $(cx_{n+1} \oplus dx_n)$ exist and belong to $\mathbf{S}$ for all choices of c and d.

For now, c and d will be left out of consideration* (by setting them both equal to 1), in order to define the simplest possible Fibonacci track recurrence. This form will be called a *Fibonacci track recurrence in S*. The sequences of objects which it generates will be called *tracks in S*, or just *tracks*.

---

*In Section 7.6, however, we shall see how 'plus-minus sequences' relate to linear (c,d)-forms in $\mathbf{Z}^3$.

Definition 1: Let the system-pair $(S, \oplus)$ be given, where S is any set of objects, and $\oplus$ is a binary relation defined on $S \times S$.

A *track in S*, relative to (a, b) and $\oplus$, is the sequence of elements in S which is generated by the *Fibonacci track recurrence*:

$$x_{n+2} = x_{n+1} \oplus x_n, \text{ with } x_0 = a, x_1 = b, \text{ and } (a, b) \in S \times S.$$

Definition 2: Two tracks in S will be deemed *equivalent* if they have the same set of elements and differ only by a cyclic displacement.

Notation: In general, a bold-face **T** will denote a track in S, and the symbol T_n its nth term. More detail may be given thus: $\mathbf{T}_\oplus(a, b)$ for the track, and $T_\oplus^{(n)}(a, b)$ for its nth element. The symbol $\mathcal{T}$ will denote the class of all (inequivalent) tracks which can be generated by the track recurrence by varying (a, b) over $S \times S$: we shall call this class *the trackset in S*.

Definition 3: The collection of sub-track triples or ('points') (x_n, x_{n+1}, x_{n+2}) which occur in the formation of a track **T** is called the *triple-set* of **T**, denoted by **Tr**. The union of the triple-sets over $\mathcal{T}$ will be denoted by $\mathcal{T}r$.

Definition 4: All tracks will be infinite, but some will be purely, and some partially, periodic. A purely periodic track of period p will be represented thus:

$$\mathbf{T} = [x_0, x_1, x_2, \cdots, x_{p-1}]_p.$$

A purely periodic track of period p has p equivalent forms, each obtained from any other by a cyclic rotation of elements. For example,

$$\mathbf{T}' = [x_1, x_2, \cdots, x_{p-1}, x_0]_p \equiv \mathbf{T}.$$

7.22 An example in Vector Geometry: rectilinear spirals

The main type of track which was studied in [29] was the *vector-product track*, defined in $S = \mathbf{Z}^3$ Thus, the binary relation was $(S, \oplus) = (\mathbf{Z}^3, \times)$, where the objects are vectors with integer coordinates and $\times$ is the vector-product operation (or 'cross-product').

The first five terms of the vector-product track in $\mathbf{Z}^3$ are:

$$\mathbf{T}_\times(\mathbf{a}, \mathbf{b}) = \mathbf{a}, \mathbf{b}, (\mathbf{b} \times \mathbf{a}), ((\mathbf{b} \times \mathbf{a}) \times \mathbf{b}), ((\mathbf{b} \times \mathbf{a}) \times \mathbf{b}) \times (\mathbf{b} \times \mathbf{a})), \cdots$$

If numerical examples of vector-product tracks are studied, the locus of their terms (i.e. points in $\mathbf{Z}^3$, joined-up sequentially by straight lines) seems to 'thrash about' wildly in the space. Analysis of the above formulae for the terms, however, shows that in general the joined-up points of a track form a rectilinear spiral, passing systematically from one to the next of three mutually orthogonal arms or directions. The points spread rapidly away from each other along these arms.

In [29] the following formulae were derived for the positions (points) represented by successive terms of the vector-product track:

Formulae: Successive terms (points) on the three arms of the spiral track are given by the following formulae for $n = 0, 1, 2, 3, \cdots$:

$$\begin{aligned}
\text{the } \mathbf{b}\text{-arm}: \quad \mathbf{T}_{3n+2} &= c^{F_{3n}} b^{F_{3n-1}-1} \mathbf{b}; \\
\text{the } \mathbf{c}\text{-arm}: \quad \mathbf{T}_{3n+3} &= c^{F_{3n+1}-1} b^{F_{3n}} \mathbf{c}; \\
\text{the } \mathbf{d}\text{-arm}: \quad \mathbf{T}_{3n+4} &= c^{F_{3n+2}-1} b^{F_{3n+1}-1} \mathbf{d}.
\end{aligned}$$

In these formulae, $\mathbf{c} = \mathbf{b} \times \mathbf{a}$ and $\mathbf{d} = b^2 \mathbf{a} - (\mathbf{a} \cdot \mathbf{b})\mathbf{b}$, $b = |\mathbf{b}|$, $c = |\mathbf{c}|$.

Note the appearances of Fibonacci numbers in the coefficients of the arm vectors. Perhaps they might have been expected to occur as exponents, since vector-products were being carried out in the recurrence, rather than additions; but that was by no means assured at the outset of the investigation. Thus the Fibonacci vector-product track is an interesting [3D]-vector figure, a spiral on three mutually orthogonal arms, which can be placed anywhere in $\mathbf{Z}^3$ by suitable choice of the initial vectors $\mathbf{a}$ and $\mathbf{b}$.

7.3 On Tracks in Groups

In Chapters 3, 4 and 6 it was shown how Fibonacci vectors (F_{n-1}, F_n, F_{n+1}) can be plotted in the honeycomb plane $x + y = z$; and that vector polygons can be drawn by joining alternating points in vector sequences by line segments. In the set of all such figures (H^2-polygons, or 'chimneys'), no polygon intersects any other polygon, and every integer-point in the honeycomb plane occurs on just one polygon. The diagram in Chapter 6 and [31] which illustrates the set of chimneys is reminiscent of a class of hyperbolas, with two asymptotes which, however, are not perpendicular to one another.

It is evident that an infinite Fibonacci vector sequence (which zig-zags within a chimney (see Ch. 6) is a Fibonacci vector track in ($\mathbf{Z}^3$, +), and that the set of all possible Fibonacci vector tracks (the trackset) exactly covers the integer-points of the Fibonacci honeycomb plane (note that these points form a group under vector addition). This is an interesting property of the trackset of Fibonacci vector tracks.

It is shown below how tracksets of finite groups can tell us interesting things about their groups, too. Similarly, the tracksets of Latin Square algebras yield useful information about their algebras.

7.31 Tracksets and Spectra of the Groups of Order 4

In this subsection it is shown how to obtain the tracksets of the two possible groups of order 4, and then their track spectra. Some of their trackset properties are presented and discussed.

First the operation tables[†] are given of the cyclic group and the Vier group, both of order four. Then their Fibonacci tracksets and associated period-spectra and identity-spectra are computed (the spectra are explained in observation (4) below, with reference to the two examples).

Fibonacci Tracksets and Spectra

Operation tables

Cyclic Group, C_4

*	e	a	b	c
e	e	a	b	c
a	a	b	c	e
b	b	c	e	a
c	c	e	a	b

Vier Group, V

*	e	a	b	c
e	e	a	b	c
a	a	e	c	b
b	b	c	e	a
c	c	b	a	e

[†]The use of operation tables was suggested by A. Cayley in *Phil. Mag. vol vii (4), 1854.* The underlying motivation for the work in this Chapter was to explore how group theory might have begun to develop, had Cayley proposed using tracksets instead.

	Cyclic Group, C_4		*Vier Group*, V

Tracksets

$$[e]_1 \qquad\qquad = \quad \tau_1$$
$$[e,a,a,b,c,a]_6 \quad = \quad \tau_2$$
$$[e,b,b]_3 \qquad\quad = \quad \tau_3$$
$$[e,c,c,b,a,c]_6 \quad = \quad \tau_4$$

$$[e]_1 \qquad\quad = \quad \sigma_1$$
$$[e,a,a]_3 \quad = \quad \sigma_2$$
$$[e,b,b]_3 \quad = \quad \sigma_3$$
$$[e,c,c]_3 \quad = \quad \sigma_4$$
$$[a,b,c]_3 \quad = \quad \sigma_5$$
$$[a,c,b]_3 \quad = \quad \sigma_6$$

$$T(C_4) \;=\; \{\tau_1,\tau_2,\tau_3,\tau_4\} \qquad\qquad T(V) \;=\; \{\sigma_1,\sigma_2,\sigma_3,\sigma_4,\sigma_5,\sigma_6\}$$

Spectra

period-spectra:

$$P(C_4) \qquad = \quad 1^1 3^1 6^2 \qquad\qquad P(V) \qquad = \quad 1^1 3^5$$
$$\text{(max. period } \pi = 6) \qquad\qquad \text{(max. period } \pi = 3)$$

identity-spectra:

$$I(C_4) \qquad = \quad 1^1 3^1 6^2 \qquad\qquad I(V) \qquad = \quad 1^1 3^3$$
$$\text{(max. period } \pi = 6) \qquad\qquad \text{(max. period } \pi = 3)$$

7.32 Observations [1]

(1) To obtain a Fibonacci trackset from an operation table (e.g. for C_4), begin with the initial pair $(x_0, x_1) = (e, e)$. Then iterate the Fibonacci track recurrence, using the table. The result is the infinite track $e, e, e, \cdots$ which has period 1: we denote this track by $[e]_1$.

Next take the pair (e, a) for initial elements, and generate the infinite track $e, a, a, b, c, a, e, a, a, b, c, a, \cdots$ which has period 6: we denote this track by $[e, a, a, b, c, a]_6$.

Repeat this process, always selecting the initial pair so that (i) it has not already appeared as a consecutive pair in an earlier track, and (ii) it is the next possible unused pair in alphabetic (i.e. dictionary) ordering of initial pairs.

(2) Track $\tau_1 = [e]_1$ obviously must occur in every group trackset, if e is the identity element of the group.

(3) In the two group tracksets shown, all the tracks are purely periodic.

(4) The Period-spectrum (P) of $C(4)$, namely $1^1\,3^1\,6^2$, signifies that there is 1 track of period 1, 1 track of period 3, and 2 tracks of period 6. Similarly, for the Vier group there is 1 track of period 1, and 5 of period 3. Thus they are *frequency distributions* of track periods, with frequencies shown as superscripts.

(5) The Identity-spectrum (I) of a group is the frequency distribution of periods of those tracks in the trackset which have the identity as an element. For the cyclic group C_4 all the tracks have one and only one identity element e, so the Period- and Identity-spectra are the same. Whereas for the Vier Group they are different, since two tracks of V do not contain an element e. Which groups have identical P- and I-spectra?

(6) Note that for both groups, $\Sigma_{\tau \in \mathcal{T}} p(\tau) \times f(\tau) = 16$, where p, f refer respectively to period and frequency of track τ,

> **Proposition:** For a group G of order g,
> $$\Sigma_{\tau \in \mathcal{T}} p(\tau) \times f(\tau) = g^2\,.$$

> **Proof:** The sum on the L.H.S. is equal to the cardinal number of the multi-set of elements occurring in all the tracks of G. But, by the method of construction of tracks, this multi-set must consist of all the elements in the group operation table, of which there are g^2. $\qquad \square$

Before discussing further the properties of tracksets, we show below the tracksets of all groups of orders 1 to 8, together with their P-spectra. It will be noted that we have used the integers of the set $\mathbf{Z}_n = \{0, 1, 2, \cdots, n-1\}$ to label all their elements: 0 is used for the identity element. A discussion on numeric labelling of elements will follow the table.

Tracksets of the fourteen groups of orders 1 to 8
(i) The eight Cyclic groups:

Gp.	Trackset	P-spectrum

Gp. **Trackset** **P-spectrum**

C_1 $[0]_1$ $P = 1^1$

C_2 $[0]_1$
$[0,1,1]_3$ $P = 1^1 3^1$

C_3 $[0]_1$
$[0,1,1,2,0,2,2,1]_8$ $P = 1^1 8^1$

C_4 $[0]_1$
$[0,1,1,2,3,1]_6$
$[0,2,2]_3$
$[0,3,3,2,1,3]_6$ $P = 1^1 3^1 6^2$

C_5 $[0]_1$
$[0,1,1,2,3,0,3,3,1,4,0,4,4,3,2,0,2,2,4,1]_{20}$
$[1,3,4,2]_4$ $P = 1^1 4^1 20^1$

C_6 $[0]_1$
$[0,1,1,2,3,5,2,1,3,4,1,5,0,5,5,4,3,1,4,$
$\qquad\qquad\qquad\qquad 5,3,2,5,1]_{24}$
$[0,2,2,4,0,4,4,2]_8$
$[0,3,3]_3$ $P = 1^1 3^1 8^1 24^1$

C_7 $[0]_1$
$[0,1,1,2,3,5,1,6,0,6,6,5,4,2,6,1]_{16}$
$[0,2,2,4,6,3,2,5,0,5,5,3,1,4,5,2]_{16}$
$[0,3,3,6,2,1,3,4,0,4,4,1,5,6,4,3]_{16}$ $P = 1^1 16^3$

C_8 $[0]_1$
$[0,1,1,2,3,5,0,5,5,2,7,1]_{12}$
$[0,2,2,4,6,2]_6$
$[0,3,3,6,1,7,0,7,7,6,5,3]_{12}$
$[0,4,4]_3$
$[0,6,6,4,2,6]_6$
$[1,3,4,7,3,2,5,7,4,3,7,2]_{12}$
$[1,4,5,1,6,7,5,4,1,5,6,3]_{12}$ $P = 1^1 3^1 6^2 12^4$

(ii) The six Non-cyclic groups:

$V =$ $[0]_1$

$C_2 \times C_2$ $[0, 1, 1]_3$, $[0, 2, 2]_3$, $[0, 3, 3]_3$

$[1, 2, 3]_3$, $[1, 3, 2]_3$ $\qquad\qquad P = 1^1 3^5$

D_3 $[0]_1$

$[0, 1, 1, 2, 0, 2, 2, 1]_8$

$[0, 3, 3]_3$, $[0, 4, 4]_3$, $[0, 5, 5]_3$

$[1, 3, 5, 2, 4, 5]_6$

$[1, 4, 3, 2, 5, 3]_6$

$[1, 5, 4, 2, 3, 4]_6$ $\qquad\qquad P = 1^1 3^3 6^3 8^1$

$C_2 \times C_2 \times C_2$ $[0]_1$

$[0, 1, 1]_3$, $[0, 2, 2]_3$, $[0, 3, 3]_3$,

$[0, 4, 4]_3$, $[0, 5, 5]_3$, $[0, 6, 6]_3$, $[0, 7, 7]_3$

$[1, 2, 6]_3$, $[1, 3, 5]_3$, $[1, 4, 7]_3$,

$[1, 5, 3]_3$, $[1, 6, 2]_3$, $[1, 7, 4]_3$,

$[2, 3, 4]_3$, $[2, 4, 3]_3$, $[2, 5, 7]_3$, $[2, 7, 5]_3$

$[3, 6, 7]_3$, $[3, 7, 6]_3$, $[4, 5, 6]_3$, $[4, 6, 5]_3$ $\quad P = 1^1 3^{21}$

$C_4 \times C_2$ $[0]_1$, $[0, 1, 1, 2, 3, 1]_6$

$[0, 2, 2]_3$, $[0, 3, 3, 2, 1, 3]_6$

$[0, 4, 4]_3$, $[0, 5, 5, 2, 7, 5]_6$

$[0, 6, 6]_3$, $[0, 7, 7, 2, 5, 7]_6$

$[1, 4, 5, 1, 6, 7]_6$

$[1, 5, 6, 3, 5, 4]_6$

$[1, 7, 4, 3, 7, 6]_6$

$[2, 4, 6]_3$, $[2, 6, 4]_3$

$[3, 4, 7, 3, 6, 5]_6$ $\qquad\qquad P = 1^1 3^5 6^8$

D_4 $[0]_1$, $[0, 1, 1, 2, 3, 1]_6$

$[0, 2, 2]_3$, $[0, 3, 3, 2, 1, 3]_6$

$[0, 4, 4]_3$, $[0, 5, 5]_3$, $[0, 6, 6]_3$, $[0, 7, 7]_3$

$[1, 4, 7, 3, 6, 7]_6$, $[1, 5, 4, 3, 7, 4]_6$

$[1, 6, 5, 3, 4, 5]_6$, $[1, 7, 6, 3, 5, 6]_6$

$[2, 4, 6]_3$, $[2, 5, 7]_3$, $[2, 6, 4]_3$, $[2, 7, 5]_3$ $\quad P = 1^1 3^9 6^6$

Q_4 $[0]_1$, $[0, 1, 1, 2, 3, 1]_6$

$[0, 2, 2]_3$, $[0, 3, 3, 2, 1, 3]_6$

$[0, 4, 4, 2, 6, 4]_6$, $[0, 5, 5, 2, 7, 5]_6$

$[0, 6, 6, 2, 4, 6]_6$, $[0, 7, 7, 2, 5, 7]_6$

$[1, 4, 7]_3$, $[1, 5, 4]_3$, $[1, 6, 5]_3$, $[1, 7, 6]_3$

$[3, 4, 5]_3$, $[3, 5, 6]_3$, $[3, 6, 7]_3$, $[3, 7, 4]_3$ $\quad P = 1^1 3^9 6^6$

7.33 Observations [2]

(1) The two groups of order 4 have different tracksets; so do the two of order 6; so do the five of order 8. It appears that if two group tracksets are equivalent, then the groups are equal. Indeed, this implication is true for finite groups, and moreover, it is two-way. In other words, two tracksets are equal (up to labelling) if and only if the two associated groups are isomorphic. The following paragraph gives an outline proof of this assertion.

It is evident from the way by which a trackset is derived from a group table, that all information about binary operations which is stored in an operation table is transferred to the tracks of the corresponding trackset. Moreover, if the trackset is given, the operation table can be recovered from the tracks, reading off its terms from successive triples in them. The set Tr of triples (which link together to form the tracks) determines both the operation table and the trackset uniquely.

(2) Likewise, the corresponding P- and I-spectra for the various groups listed in (1) are different. It is evident that a trackset spectrum is an invariant of its group. We do not know whether two groups may have the same spectrum: I doubt whether this could be so. We have examples of two different Latin-square algebras which have the same P-spectra; but they are not groups.

(3) The numeric labelling of the objects was not done arbitrarily, although it could have been: the integers (symbols) in $\mathbf{Z}_n$ are merely substitutes for symbols such as $e, a, b, c, \cdots$ We shall not spell out my labelling method, except for the case of the cyclic groups: for those we assigned the integer symbols from $\mathbf{Z}_n$ so that the resulting tracksets were equivalent to the tracksets obtainable from $(\mathbf{Z}_n, \mid \times \mid_n)$; these systems are well-known to be isomorphic to the cyclic groups.

(4) Since the early 1960s Fibonacci sequences modulo n (which are sequences in C_n) have been much studied [1]. Hence most, perhaps all, of the properties given here about tracksets of cyclic groups may be already well-known. However, trackset methods can be used with any finite group of order n, using the elements of Z_n to label the group's elements. The tracksets will then indicate structural properties of the group, particularly those which are invariant to changes in object-labellings. The period-spectrum of $\mathcal{T}$ is one such property. Indeed, one can go on to handle the tracks and tracksets as if they were composed of integers (which in general they are

not) provided one keeps in mind that the integers are just symbols for group elements, and takes care not to assume number attributes for the elements which they don't possess, for example ordering properties.

An interesting question to ask is: How does a permutation of the elements of $\mathbf{Z}_n$, when applied to all the elements of a group and its trackset. affect the tracket's reflection of the group's properties? The answer is: Not at all! The trackset remains the same, but with different symbols for its elements; hence it tells the same story about the group, whose symbols have changed in the same way. In particular, its period-spectrum remains exactly the same.

Some tracksets are even invariant under all permutations applied to the set of their elements (other than their identity). This would seem to be a strong invariance property. For example, keeping 0 fixed, permuting the elements of $\mathbf{Z}_n \backslash \{0\}$ in any possible way leaves the tracksets of C_1, C_2, C_3 and the Vier group V unchanged.

7.4 Equivalence of Tracksets and Groups

Before examining properties of tracksets further, we shall spell out which properties a trackset must have for it to be equivalent to a group. Here 'equivalence' is used in the sense that a group may be derived, uniquely, from a trackset, and vice-versa.

Let $\mathcal{T}$ be a set of tracks (i.e. a trackset) in $(S, \oplus)$, with $|S| = n$; and let $\mathcal{T}r$ be its triple-set union (see Section 2, Definition 3). Without loss of generality, we shall let $S = \mathbf{Z}_n$.

7.41 Equivalence of trackset with a group

$\mathcal{T}$ is equivalent to a finite group of order n if:

G1: With every ordered pair of elements a, b in $\mathbf{Z}_n$ there is associated a unique triple (a, b, c) in $\mathcal{T}r$;

G2: For each $x \in \mathbf{Z}_n \backslash \{0\}$, both $(0, x, x)$ and $(x, 0, x)$ are in $\mathcal{T}r$. (0 is the identity element of the group.)

G3: For each $a \in \mathbf{Z}_n$, there is an $x \in \mathbf{Z}_n$ with both $(a, x, 0)$ and $(x, a, 0)$ being in $\mathcal{T}r$. (x is called the inverse of a.)

G4: Let a, b, c be any three elements of $\mathbf{Z}_n$; then, by G1, there are (a, b, r) and (b, c, s) in $\mathcal{T}r$. There exists an $x \in \mathbf{Z}_n$ such that both (a, s, x) and (r, c, x) are in $\mathcal{T}r$.

It may be seen that these postulates correspond directly with those usually given as axioms for a group, viz.: G1 (closure), G2 (existence of an identity element), G3 (existence of unique inverses) and G4 (associative law holds).

From G1, there are n^2 triples in the trackset; and each triple (a, b, c) corresponds to a relation $a \oplus b = c$. Clearly, the group addition table is obtainable from the trackset, and vice-versa.

We can drop the requirement for finiteness, and allow infinite tracks and tracksets. The same postulates then ensure equivalence of an infinite trackset with an infinite group.

7.5   Some Operations with Tracksets

In this Section, we show first how useful tracksets are for determining whether a binary relation is a group relation, and for discovering subgroups by checking for inclusion of their tracksets within that of a larger group trackset. Then we study certain properties of tracksets; in particular we look at the period spectra of cyclic groups.

7.51 Checking if a binary relation is a group relation

If we are given a binary operation on a finite set, we can determine its trackset directly, without first writing out the operating table. Then we can compare this trackset with those listed in Section 3. The following example demonstrates the procedure.

Example 1: (quadratic residues modulo 11)

It is easily checked that if $n \in \mathbf{N}$, and r is the residue of n^2 modulo

11, then $r \in \{1, 3, 4, 5, 9\} \equiv M_{11}$. We show next that with multiplication modulo 11, M_{11} is a group.

The trackset, determined directly, is:

$$\begin{aligned}
\tau_1 &= [1]_1 \\
\tau_2 &= [1, 3, 3, 9, 5, 1, 5, 5, 3, 4, 1, 4, 4, 5, 9, 1, 9, 9, 4, 3]_{20} \\
\tau_3 &= [3, 5, 4, 9]_4
\end{aligned}$$

N.B. We began with $|1 \times 1|_{11} = 1$; then $|1 \times 3|_{11} = 3$, and so on, until after 4,3 the sequence in τ_2 repeated itself. Then we checked that 3,5 was the first pair of elements not included in the first two tracks, so track τ_3 began with those. Since three tracks use up $5^2 = 25$ elements, and $|M|_{11}$ has 5 elements, we know that the trackset is complete.

The only group of order 5 is C_5. Comparing the M_{11} trackset with the C_5 trackset, we see that their period spectra are the same (viz. $P = 1^1 4^1 20^1$): and a quick check shows that the mapping $(0, 1, 2, 3, 4) \leftrightarrow (1, 3, 9, 5, 4)$ carries one trackset into the other. Hence M_{11} is a cyclic group of order 5, under multiplication modulo 11.

7.52 Determining subgroups

The next example shows how to determine subgroups from tracksets.

Example 2:

If we are given the trackset of a group, we can determine from its P-spectrum, and a standard list of spectra of smaller groups, which sub-tracksets are possible candidates for subgroups.

For example, the Vier group has P-spectrum $1^1 3^5$. We know that the trivial subgroup (spectrum 1^1) is a subgroup. And since C_2 has spectrum $1^1 3^1$, there are five candidates for subgroups of V. The trackset for C_2 is $\{ [0]_1 , [0, 1, 1]_3 \}$; and from V we see that the following three sub-tracksets are all of this form:

$$\{ [0]_1 , [0, 1, 1]_3 \}, \{ [0]_1 , [0, 2, 2]_3 \}, \{ [0]_1 , [0, 3, 3]_3 \}.$$

Thus the subgroups of V are $< 0 >$, $< 0, 1 >$, $< 0, 2 >$ and $< 0, 3 >$.

A slightly more complex example is the quarternion group Q_4. Its P-spectrum (see the table in Section 3) is $P = 1^1 3^9 6^6$, which indicates that $C_1(1^1)$, $C_2(1^1 3^1)$, $C_4(1^1 3^1 6^2)$ and $V_4(1^1 3^5)$ are subgroup candidates.

Looking at the tracks of Q_4, we quickly determine that $< 0 >$, $< 0, 2 >$, $< 0, 1, 2, 3 >$ and $< 0, 2, 4, 6 >$ are the only subgroups. The Vier group, V_4, cannot be a subgroup, since Q_4 doesn't have three tracks of period 3 which include the identity element.

As an aside, we comment that in an early paper on groups, circa 1855, Arthur Cayley announced that he had found three groups of order 6. In fact there are only two, D_3 and C_6. Had he done this simple trackset check, the P-spectra would have shown him that two of his three were isomorphic.

7.53 Properties of trackset-spectra of cyclic groups

We have computed the tracksets of the first 127 cyclic groups, and there follow a few observations upon them. It is likely that many of the points made will be well-known, so we shall be brief, and not give any proofs.

(1): The track beginning $[0, 1, 1, ...]$ in each trackset has maximal period, say π^*. It is observed in all P-spectra that all track periods divide the maximal period π^* evenly.

(2): In [1], [7] it is stated that the period for an odd prime modulo p is related to the entry point $Z(p)$ in the Fibonacci sequence $F(1, 1)$, as follows. The sequence modulo p has period which is respectively $4 * Z(p), 2 * Z(p)$, or $Z(p)$ according as $Z(p)$ is odd, $|Z(p)|_4 = 0$ or $|Z(p)|_4 = 2$.

For example, the entry point for $p = 5$ is $Z(5) = 5$, since $F_5 = 5$. Then, since $Z(5)$ is odd, the period is $4 * Z(5) = 20$. It may be checked from our table of tracksets, that C_5 has a track of period 20, which begins $[0, 1, 1, ...]$.

(3): Whilst the theories and tables of [1] and [7] do much to explain the properties of P-spectra of cyclic groups, those for non-cyclic groups are not so easily explained, although it is clear that entry-points still play a crucial role.

(4): By inspection of the first hundred or so of tracksets of cyclic groups with prime order, it appears that their P-spectra take certain simple forms. They are expressible, both their periods and period-frequencies, in terms of p, $(p - 1)$ or $(p + 1)$, as shown in the tables below; and they are related to the entry points of p in the basic Fibonacci sequence.

One-period C_p tracksets

Labels	Forms	Values of p	Entry points Z(p)	Z(p) relations	$|Z|_4$
A	$(p+1)^{(p-1)}$	2	3	$Z = p+1$	$|Z|_4 = 3$
B	$\frac{2}{3}(p+1)^{\frac{3}{2}(p-1)}$	47, 107	16, 36	$Z = \frac{1}{3}(p+1)$	$|Z|_4 = 0$
		103	19	$Z = \frac{1}{6}(p+1)$	$|Z|_4 = 3$
C	$2(p+1)^{\frac{1}{2}(p-1)}$	3, 7, 23, 43	4, 8, 24, 44	$Z = \frac{1}{1}(p+1)$	$|Z|_4 = 0$
		67, 83, 103, 127	68, 84, 104, 128		
		17, 73, 97	9, 37, 49	$Z = \frac{1}{2}(p+1)$	$|Z|_4 = 1$
		13, 37, 53	7, 19, 27	$Z = \frac{1}{2}(p+1)$	$|Z|_4 = 3$
D	$(p-1)^{(p+1)}$	41	20	$Z = \frac{1}{2}(p-1)$	$|Z|_4 = 0$
		61, 109	15, 27	$Z = \frac{1}{4}(p-1)$	$|Z|_4 = 3$
E	$\frac{1}{2}(p-1)^{2(p+1)}$	89	11	$Z = \frac{1}{8}(p-1)$	$|Z|_4 = 3$

Two-period C_p tracksets

Labels	Forms	Values of p	Entry points Z(p)	Z(p) relations	$|Z|_4$
F	$(p-1), p(p-1)$	5	5	$Z = p$	$|Z|_4 = 1$
G	$(\frac{p-1}{2})^2, (p-1)^p$	11, 19, 31	10, 18, 30	$Z = p-1$	$|Z|_4 = 2$
		59, 71, 79	58, 70, 78		
H	$(\frac{p-1}{4})^4, (\frac{p-1}{2})^{2p}$	29, 101	14, 50	$Z = \frac{1}{2}(p-1)$	$|Z|_4 = 2$

First there is a dichotomy. Disregarding the 1-period identity track, which occurs in every spectrum, there are either one period, or two periods, in the P-spectrum of C_p if p is prime. The above two tables show the trackset forms for $p = 2, 3, 5, ..., 127$.

(5): Within the tables, it may be observed that when $|p|_{10}$ is 3 or 7, the trackset periods are all of the form $a(p+1)$, with frequencies of form $b(p-1)$, with $a, b \in \mathbf{Q}^+$: and $Z(p)|(p+1)$.
Whereas, when $|p|_{10}$ is 1 or 9, the periods are of form $a(p-1)$ and $b(p-1)$, with frequencies of 2^n and $2^{n-1}p$ respectively: and $Z(p)|(p-1)$.

The above observations lead us to make the following conjectures[‡].

[‡]Dr. L. Somer (The C. U. A.) published papers on Fibonacci sequences modulo p in the Fibonacci Quarterly, in the 1980s. In a recent conversation with this author, he suggested that his results would settle these conjectures. He thought, however, that the

7.54 Conjectures on tracksets of cyclic groups:

In the following, we omit primes 2 and 5 from consideration; and we do not count the track of period 1 which occurs in every group-trackset.

(1) Every trackset P-spectrum has one or two period-values.

(2) For those tracksets with one period-value, the P-spectrum has form

either (i) period $= a(p+1)$ with frequency $= 1/a(p-1)$,

or (ii) period $= b(p-1)$ with frequency $= 1/b(p+1)$,

with $a,\ b \in \mathbf{Q}^+$.

An infinity of values of a, b will occur as $p \to \infty$.

(3) For those tracksets with two period-values (p_1 and p_2) the P-spectrum has form:

$$p_1{}^{f_1},\ p_2{}^{f_2} = \left(\frac{p-1}{2^n}\right)^{2^n},\ \left(\frac{p-1}{2^{(n-1)}}\right)^{2^{n-1}p}, \text{ where } n \in \mathbf{N}.$$

An infinity of values of n will occur as $p \to \infty$.

(4) In all cases, (i) if $|p|_{10} = 1$ or 9, then $Z(p)|p-1$; whereas (ii), if $|p|_{10} = 3$ or 7, then $Z(p)|p+1$.

(5) (i) In cases (2), for each form $|p|_{10}$ is either (1 or 9) or else it is (3 or 7).

(ii) In cases (3), for every form $|p|_{10}$ is (1 or 9).

(6) (i) In cases (2), for every form of P-spectra examples will exist such that $Z(p)_4 = 0, 1$, or 3.

(ii) In cases (3), every form of P-spectra will have $|Z(p)|_4 = 2$; and $Z(p) = (p-1)/n$ with $n \in \mathbf{N}$.

In the next Section, some tracksets are shown which are equivalent to the operation tables of what the author calls 'plus-minus algebras'. These algebras are not generally groups; but they have Latin square operation tables, and so in this respect their tracksets can be compared with those of cyclic groups, studied above. Furthermore, the 'plus–minus' sequences are simple generalisations of the Fibonacci sequences modulo p, which adds interest to their study in this context.

conjectures were interesting in the context of tracksets. We are indebted to Dr. Somer for his information and comments.

7.6 Plus-Minus Sequences Modulo p

7.61 Definitions

A 'plus-minus' sequence of integers is generated by the recurrence given in Definition 2 below. It will be observed that this is, in fact, a Fibonacci track form in $(\mathbf{Z}, [\pm]_m)$, where $[\pm]_m$ is the binary operation described in Definition 1.

Definition 1: (the plus-minus operation)

Let $[\pm]_m$ be a binary operation on elements a, b, which involves a succession of additions or subtractions as determined by a particular (given) string of plus and minus signs, of length m.

For example, if m is 3 and the given string is $(+ - +)$, then $a[\pm]_3 b = a(+ - +)b = 2a + b$, obtained thus: first we get, applying the plus sign, $a, b \to (a + b)$; then, applying the minus sign, $(a+b) - b \to a$; finally, applying the plus sign gives $a + (a + b) \to 2a + b$.

Definition 2: (the plus-minus recurrence form)

Let S_1 and S_2 be given integers. Then the sequence $\{S_n\}$ of integers obtained by applying the recurrence $S_{n+2} = S_{n+1} [\pm]_m S_n$ is the *Fibonacci plus-minus track* with the given initial values and the given plus-minus string of length m.

7.62 Equivalence with a linear Fibonacci recurrence

Following the example given in Definition 1, it is not difficult to see that whatever $[\pm]_m$ operation is given, the result will be a linear form $ca + db$; and that g.c.d.$(c, d) = 1$ if $c \neq 0$ and $d \neq 0$.

Thus each plus-minus recurrence form is equivalent to a Fibonacci recurrence of the type: $S_{n+2} = cS_{n+1} + dS_n$ for some pair (c, d).

It remains for us to find an algorithm for determining the pair (c, d) for each given string $[\pm]_m$. The tree-diagram below gives a neat solution to this problem.

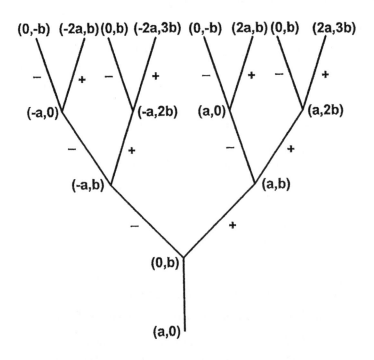

The tree for plus-minus recurrence forms

The ordered pairs on the nodes of the tree indicate the required forms. For example, using again $[\pm]_3 = (+ - +)$, proceeding from the lowest branching node along the three branches labelled $+, -, +$ takes us to the node labelled $(2a, b)$. This means that $S_{n+2} = S_{n+1}[\pm]_3 S_n = 2S_{n+1} + 1S_n$.

This tree, and its relationships with plus-minus recurrences, has many interesting properties. We have space to mention only a few.

(i) Each pair (c, d) occurs on an infinity of nodes: thus there is an infinity of plus-minus strings which correspond to the same (c, d) recurrence.

(ii) If we set any plus-minus recurrence sequence in vector form[§] we get a sequence of vectors in the plane $z = cx + dy$: this is an *inherent sequence* in that plane (see Ch. 2 and Ch. 6).

[§]To do this, replace each triple (S_{n-1}, S_n, S_{n+1}) in the sequence by the vector $\mathbf{S_n}$

(iii) The pairs (c, d) from the tree's nodes can be put in 1-1 correspondence with those on the trees used to model the modular group in paper [28]. A sub-tree provides all pairs corresponding to rational fractions $c/d \in \mathbf{Q}$ (see the rational number tree in [26]).

For the purposes of this book, we wish only to present trackset-spectra of a few of the Latin-square algebras derivable from plus-minus sequences modulo p. For each choice of $(p : (c, d))$ there is a an algebra on $\{0, 1, 2, ..., p-1\}$ which has a Latin-square operation table (if $c, d \neq 0$).

In the table below are shown the resulting P-spectra for seven different choices of (c, d) and with $p = 7, 11, ..., 41$.

P − spectra $p\backslash(c,d)$	Cyclic (1,1)	(1,2)	Pell(2) (2,1)	(2,3)	(1,3)	Pell(3) (3,1)	(3,2)
7	16^3	$3^2 6^7$	$2^3 3^2 6^6$	48^1	16^3	24^2	$2^3 6^7$
11	$5^2 10^{11}$	5^{24}	$2^5 10^{11}$	30^4	8^{15}	120^1	$2^5 5^2 10^{10}$
13	28^6	6^{28}	$2^6 12^{13}$	$3^4 12^{13}$	$4^3 52^3$	$12^1 156^1$	$2^6 3^4 6^{24}$
17	36^8	16^{18}	$2^8 8^{34}$	$16^1 272^1$	16^{18}	$4^4 16^{17}$	$2^8 16^{17}$
19	$9^2 18^{19}$	40^9	$2^9 18^{19}$	$6^3 18^{19}$	40^9	90^4	$2^9 18^{19}$
23	48^{11}	$11^2 22^{23}$	$2^{11} 11^2$	176^3	$11^2 22^{23}$	$11^2 22^{23}$	$2^{11} 11^2 22^{22}$
29	$7^4 14^{28}$	20^{42}	$2^{14} 28^{29}$	840^1	28^{30}	$7^4 28^{29}$	$2^{14} 28^{29}$
31	$15^2 30^{31}$	$15^2 30^{31}$	$2^{15} 5^6$	320^3	64^{15}	240^4	$2^{15} 30^{31}$
37	76^{18}	76^{18}	$2^{18} 36^{37}$	1368^1	76^{18}	171^8	$2^{18} 18^{74}$
41	40^{42}	$5^8 10^{164}$	$2^{10} 20^{82}$	280^6	28^{60}	336^5	$2^{20} 8^{205}$

Trackset P-spectra for various values of $p\backslash(c,d)$

7.63 Comments and comparisons

There are many interesting comments and comparisons that can be made about the spectra shown above. We have space for only a few brief remarks.

(i) The first column is of the cyclic groups, treated above. The spectra are not all unique; for example, 16^3 occurs in columns $(1, 1)$ and $(1, 3)$ when $p = 7$.

(ii) Tracksets with periods of frequency 1 are rare, and therefore of interest. For $(c, d) = (2, 3)$ they occur three times, when $p = 7, 29$ and 37. For these algebras there is just one track, passing through all of the $p^2 - 1$ tabled elements, before repeating. Will there be more such tracksets in the column headed $(2, 3)$ as $p \to \infty$? An infinite number of them? What rule causes them to occur?

(iii) Occurrences of tracksets having three different periods are also rare. Three occur in the column headed $(3, 2)$, and one in column $(2, 1)$. They all fit the formula $2^{\left(\frac{p-1}{2}\right)}\left(\frac{p-1}{2}\right)^2 (p-1)^{(p-1)}$. What is the underlying reason for this?

(iv) Three trackset-spectra in column $(1, 2)$ are equal to the three corresponding ones (when $p = 17, 23, 37$) in column $(1, 3)$. Why does this happen? We hope that further study will shed light on this question.

Next we show briefly how tracksets of the vector forms of sequences treated above may be said to correspond to knots in $\mathbf{Z}^3$.

7.7 A Knot from a Trackset

If we follow the techniques of Ch. 1 and papers [30] and [31], we can convert the tracks of finite groups and plus-minus algebras into sequences of vectors in $\mathbf{Z}^3$. If we then 'join up consecutive points' with line-segments, each track becomes a closed polygon in $\mathbf{Z}^3$. With care, we can define this process (making infinitessimal adjustments in coordinates where necessary) so that no track polygon is self-intersecting; and, further, that no two track polygons have common points. Thus, every track (except $[0]_1$) is topologically equivalent to a knot. We define the identity track $[0]_1$ (now regarded as an infinite sequence of vectors $(0, 0, 0)$) to be equivalent to a degenerate null knot.

Thus, if a Latin-square algebra has a trackset of ν tracks, its corresponding set of vector polygons (with appropriately well-defined coordinate adjustments to avoid intersections) is topologically equivalent to a set of ν knots; that is, to a ν-link in $\mathbf{Z}^3$.

Study of these links reveals that each track (i.e. its vector polygon) moves within and between a stack of planes $z = x + y + m$, $m \in \mathbf{Z}$, all parallel to the Fibonacci honeycomb plane $z = x + y$ (see Ch. 3). In particular, for the cyclic groups modulo p, the planes are just two, viz. $z = x + y$ and $z = x + y - p$.

We give examples only of those knots which arise from the first five finite groups, of orders 1 through 4. The cyclic groups have knots as follows:

C_1 is a degenerate null knot (as defined).

C_2 has a knot which is the disjoint union of the C_1 knot (all group knots have this degenerate knot as a component) and the null knot obtained from the triangle on points $(0, 1, 1), (1, 1, 0), (1, 0, 1)$. Note that the first and third points are in plane $x + y = z$, and the second point is in plane $x + y - 2 = z$.

C_3 has only two tracks. One gives the degenerate knot. The other is $[0, 1, 1, 2, 0, 2, 2, 1]_8$ which provides eight points, of which 5 are in plane $x + y = z$ and 3 are in $x + y - 3 = z$. The vector polygon intersects itself at one point at which we change x, y, z to $x, y, -\delta z$. The closed polygon is then equivalent to a null knot.

C_4 has four tracks. One provides the degenerate knot, and the knot resulting from the other three is a 3-link which is commonly known as the Russian Wedding Rings. In the topological knot tables it is the link 6_3^3. Its diagram is shown below.

The only non-cyclic group with $n \leq 4$ is the Vier group:

V_4, the Vier group, has five tracks other than $[0]_1$, each of which has a 3-point vector polygon. Their points lie in a stack of four parallel honeycomb planes, with parameters $m = 0, -2, -4, -6$. The equivalent knot is the disjoint union of a degenerate knot, a null knot, and a chain of four linked rings. Its diagram is shown below.

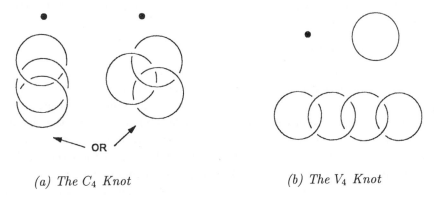

(a) *The C_4 Knot* (b) *The V_4 Knot*

7.8 On Aesthetics and Applicability of Tracksets

In this Chapter the concept of a *Fibonacci trackset*, denoted by the symbol $\mathcal{T}$, has been defined and illustrated. The author has attempted to show not only that tracksets are elegant mathematical objects, having many interesting properties that beg for deeper study, but also that they are useful tools for investigating a wide variety of algebraic structures. Indeed, he has shown how tracksets actually *define* binary operation algebras, and in themselves display, or hint at, many of their properties.

To review examples of trackset use, and to discuss associated aesthetic values, let us consider first the tracksets of the cyclic groups (see the table in Section 3). We wish to do this in the spirit of J. P. King's plea [16, p. 181], in his book *The Art of Mathematics.* He says that mathematicians should attempt to assess¶ the aesthetic values of any mathematical concepts which they create or discover. He suggests that we use two principles as standards by which the aesthetic quality of a mathematical notion can be gauged, namely the principles of *minimal completeness* and of *maximal applicability.* The closer the new notion meets these two standards, the higher is its aesthetic quality.

For *tracksets* (say T) to score highly on these tests, it must be demonstrated [quote] (1) that T contains within itself all properties necessary to fulfil its mathematical mission, with T containing no extraneous properties (i.e. minimal completeness), and (2) that T contains properties which are widely applicable to mathematical notions other than T (i.e. maximal applicability).

The following paragraphs give evidence for a good rating of T against both of these standards.

Minimal completeness

Tracksets are exactly equivalent to operation tables; the one can be derived from the other, and vice versa. Thus a trackset completely defines a binary algebra. Its 'mathematical mission' is to enable lists of properties of the algebra to be discovered; that is, to develop and analyse the algebra. This it can do, or at least it provides a complete basis for doing so. It is minimal in the sense that with any less information than the trackset contains, the binary algebra would not be completely defined.

[It is conceded that there are other ways of presenting algebras, each of which provides its own benefits for further study of its algebra: for example, representations of groups by means of generators and relations.]

¶The noted American painter Robert Henri has said of 'telling': "Low art is just telling things, for example 'There is the night.'; High art gives the feel of the night."

The next two paragraphs compare and contrast the use of tracksets and operation tables for studying groups.

The operation table of a cyclic group is merely a square matrix with a top row of $\{0, 1, 2, \cdots, x_{n-1}\}$, followed by $n-1$ rows which are progressively cycled permutations of the top one. The result is an $n \times n$ matrix having all forward (upward) diagonals sporting constant elements: a serene, simple, block of n^2 integers – pretty but bland. Moreover, with regard to the whole class of cyclic groups, if you've seen one such table you have seen them all (likewise for the generator/relation presentations). There is no hint in them that cyclic groups have different properties amongst themselves, with interesting (sometimes surprising) actions and subgroups to explore.

By contrast, even a cursory examination of the tracksets of groups (cyclic and non-cyclic; and also of other binary operation structures) reveals all manner of patterns worthy of study. Subgroups can be readily discovered as subsets of tracks, and checked against smaller tracksets; and the period- and identity-spectra vary widely, with interesting properties presenting themselves for study; the periodic behaviour of elements within tracks, related to Fibonacci sequence entry-point theory, also holds much fascination. If work were to be done to understand and classify trackset properties, then new developments in the algebraic structures which they define would be bound to follow. The contrasts in aesthetic appeals of operation tables versus tracksets seem to the author to be 'static and bland' versus 'dynamic, vigorous and revealing'.

Maximal applicability

With regard to standard (2), the Venn diagram below shows adequately that the notion $\mathcal{T}$ has a huge range of 'mathematical applicability'. It was a pleasant surprise to the author to realise that Fibonacci tracksets (i.e. $\mathcal{T}$) could be used to underpin – indeed define – every binary algebraic structure.

A good case has surely been made for tracksets to be highly rated by King's two aesthetic-value principles.

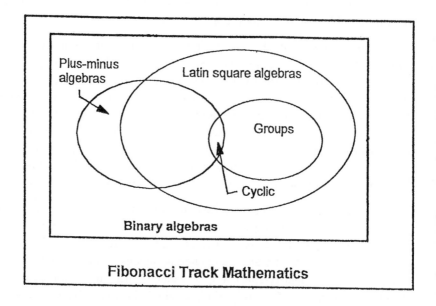

*Figure 1. Venn diagram showing the embrace
of Fibonacci Trackset mathematics*

Bibliography

Part B, Section 1

Fibonacci Vector Geometry

1. ALFRED, Brother U. 1965: *Tables of Fibonacci entry points (for primes 2 through 99,991)*. Pub. by The Fibonacci Association.
2. ARKIN, J., Arney, D. C., Bergum, G. E., Burr, S. A., Porter, B. J. 1989: Recurring sequence tiling. *The Fibonacci Quarterly* 27.4, 323–332.
3. —1990: Tiling the kth power of a power series. *The Fibonacci Quarterly* 28.3, 266–272.
4. BERGUM, G. E. 1984: Addenda to geometry of a generalized Simson's formula. *The Fibonacci Quarterly* 22.1, 22–28.
5. BROUSSEAU, A. 1972: Fibonacci numbers and geometry. *The Fibonacci Quarterly* 10.3, 303–318,323.
6. CLARKE, J. H. 1964: Linear diophantine equations applied to modular coordination. *Australian Journal of Applied Science* 15.4, 345–348.
7. FIELDER, D. C. and Bruckman P. S. 1995: *Fibonacci entry points and periods for primes 100,003 through 415,993*. Pub. by The Fibonacci Association.
8. GREEN, S. L., 1950: *Algebraic solid geometry*. Cambridge University Press.
9. HERDA, H. 1981: Tiling the plane with incongruent regular polygons. *The Fibonacci Quarterly* 19.5, 437–439.
10. HILTON, P. and Pedersen, J. 1994: A note on a geometric property of Fibonacci numbers. *The Fibonacci Quarterly* 32.5, 386–388.
11. HOGGATT Jr., V. E. and Alladi, Krishnaswami, 1975: Generalized Fibonacci tiling. *The Fibonacci Quarterly* 13.3, 137–145.
12. HOGGATT Jr.,V. E. 1979: *Fibonacci and Lucas numbers*. The Fibonacci Association; 56–57.

13. HOLDEN, H. L. 1975: Fibonacci tiles. *The Fibonacci Quarterly* 13.1, 45–49.

14. HORADAM, A. F. 1961: A generalized Fibonacci sequence. *American Mathematical Monthly* 68, 455–459.

15. —1982: Geometry of a generalized Simson's formula. *The Fibonacci Quarterly* 20.2, 164-68.

16. KING, J. P. 1992: *The art of mathematics.* Plenum Press.

17. KLARNER, D. A. and Pollack, J. 1980: Domino tilings of rectangles with fixed width. *Discrete Mathematics* 32.1, 53–57.

18. —1981: The ubiquitous rational sequence. *The Fibonacci Quarterly* 19.3, 219–228.

19. OKABE, A., Boots, B. and Sugihara K. 1992: *Spatial tessellations: concepts and applications of Voronoi diagrams.* Chichester: J. Wiley.

20. PAGE, W. and Sastry, K. R. S. 1992: Area-bisecting polygonal paths. *The Fibonacci Quarterly* 30.3, 263–273.

21. READ, R. C. 1980: A note on tiling rectangles with dominoes. *The Fibonacci Quarterly* 18.1, 24–27.

22. RUCKER, R. 1997: *Infinity and the mind.* Penguin Books, 36–38.

23. STEIN, S. and Szabo, S. 1994: *Algebra and tiling.* Carus Mathematical Monographs, No. 25, Mathematical Association of America.

24. TURNER, J. C. 1988: On polyominoes and feudominoes. *The Fibonacci Quarterly* 26.3, 205–218.

25. TURNER, J. C. 1988: Fibonacci word patterns and binary sequences. *The Fibonacci Quarterly* 26.3, 233–246.

26. TURNER, J. C. 1990: Three number trees — their growth rules and related number properties. In Bergum, G. E. et al. (eds.) *Applications of Fibonacci Numbers* 3. Kluwer Academic Publishers, 335–50.

27. TURNER, J. C. and Schaake, A.G. 1993: The elements of enteger geometry. In G. E. Bergum et al. (eds.) *Applications of Fibonacci Numbers* 5, A. P. Kluwer, 569–583.

28. TURNER J. C. and Schaake, 1996: On a Model of the Modular Group. In Bergum, G. E. et al. (eds.), *Applications of Fibonacci Numbers* 6. Kluwer Academic Publishers, 487–504.

29. TURNER, J. C. 1998: The Fibonacci track form, with applications in Fibonacci vector geometry. In dedicatory volume, to J. C. Turner (70th birthday), *Notes on Number Theory and Discrete Mathematics*, Bulgarian Academy of Sciences, 4, 136–147.

30. TURNER, J. C. and Shannon, A.G. 1998: Introduction to a Fibonacci geometry. In G. E. Bergum et al. (eds.) *Applications of Fibonacci Numbers* 7, A. P. Kluwer, 435–448.

31. TURNER, J. C. 1999: On vector sequence recurrence equations in

Fibonacci vector geometry. In Bergum, G. E. et al. (eds.), *Applications of Fibonacci Numbers* **8** Autumn. Kluwer Academic Publishers, 353–68.

32. TURNER, J. C. and Shannon, A. G. May, 2000: On Fibonacci sequences, geometry, and the m-square equation. *The Fibonacci Quarterly* 38.2, 98–103.

33. VASSILEV, V. 1995: Cellular automata and transitions. In S. Shtrakov and Mirchev (eds.) *Discrete Mathematics and Applications*, Neofit Rilski University, Blagoevgrad, 196–207.

34. WU, T. C. 1983: Counting the profiles in domino tiling. *The Fibonacci Quarterly* 21.4, 302–304.

PART B: GOLDPOINT GEOMETRY

SECTION 2

GOLDPOINT GEOMETRY

Vassia Atanassova and John Turner

'Goldpoint' is a word coined by Turner to designate a point of golden section in a line segment. Goldpoints arise naturally in certain geometric figures, and it is always of interest to discover them, and to find out why they are there in the figures. Such studies have often been recorded by mathematicians since the ancient times of the great Greek geometers. In this Section we explore new possibilities for introducing goldpoints into figures, making constructions which involve them, and analysing the consequences.

Turner has dubbed the study of goldpoints in geometric figures, whether discovered or introduced, as 'goldpoint geometry'. Many interesting examples of such studies are given in the next seven chapters.

Chapter 1

On Goldpoints and Golden-Mean Constructions

1.1 Introduction

In [1] Turner introduced a notion which he called *Goldpoint Geometry*. It consists of the study of geometric figures into which golden-mean points have been constructed or introduced. Such points he defined to be 'goldpoints'.

In fact, goldpoint geometry began with a Christmas puzzle. In late 1996, Turner sent the Atanassov family a Christmas card, on which he had drawn the star diagram shown below, and had set a puzzle about its crossing points for them to attempt.

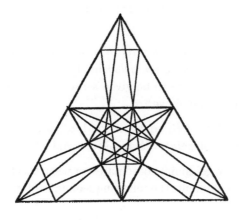

The Fibonacci Star

He included also a set of equilateral triangles, made from card and marked with golden section points; various jig-saw problems with these cards were suggested. The Atassanovs became interested in these problems, and subsequent developments with related geometric and combinatoric themes led to results which are to be described in this Section. *Christmas puzzle:* Find how many points in this Star are interior golden section points of line segments, given that the figure is based on four equilateral triangles, and that the two boundary points immediately below the apex are golden section points of the top triangle's sides.

In subsection 1.6 we discuss this puzzle in detail. Before that, however, we define 'goldpoints' and then introduce the reader to many simple results and examples of studies in goldpoint geometry.

Definition of goldpoints

In general, P is an interior goldpoint with respect to a line-segment AB if P is an interior golden-mean of the line-segment. There are two candidates for the position of a goldpoint in AB.
There are also two exterior goldpoints relative to AB.

We can achieve a simple definition of all four goldpoints if we assign a sense $(\pm)$ to segments in the line of AB, according as they are traversed (or described) in the direction $A \rightarrow B$ $(+)$, or in the direction $B \rightarrow A$ $(-)$.

Definition:
(i) If AB is a line-segment, and P is a point in the line of AB such that $|AP : PB|$ equals α or $1/\alpha$ (where $\alpha = (1 + \sqrt{5})/2$) then P is a *goldpoint* with respect to AB.

(ii) A goldpoint is an *interior goldpoint* if AP/PB is positive, and an *exterior goldpoint* if AP/PB is negative. (AP and PB are to be given senses $(\pm)$ as described in the paragraph above.)

(iii) If $|AP : PB|$ equals α^i or $1/\alpha^i$, we shall call P an *ith-order goldpoint* with respect to AB.

When calculating goldpoint coordinates, or when checking to see whether a given point is a goldpoint, the following lemma is often most useful.

Lemma 1: The interior goldpoints with respect to two points $\underline{A}$ and $\underline{B}$ are $\underline{G} = 1/\alpha\underline{A} + 1/\alpha^2\underline{B}$ and $\underline{H} = 1/\alpha^2\underline{A} + 1/\alpha\underline{B}$.

Proof: This follows from the fact that the weights $1/\alpha$ and $1/\alpha^2$ sum to 1, and form ratios $\alpha : 1$ or $1 : \alpha$. In calculations, use is often made of the identitites $1/\alpha = \alpha - 1$ and $1/\alpha^2 = 2 - \alpha$. $\qquad\square$

Whenever we are studying only ratios, we may assume the length of the line-segment AB to be $|AB| = 1$. Then $|AG| = 1/\alpha^2$, $|GH| = 1/\alpha^3$ and $|HB| = 1/\alpha^2$. The situation is shown in Figure 1 below.

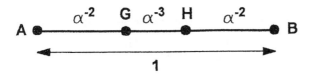

Figure 1. Unit segment AB, and its goldpoints

The following lemma is about cross-ratios in AB, with respect to its goldpoints. It follows directly from formulae given in [6]. A second lemma is given below it, which links goldpoints with harmonic ratios. These lemmas indicate that interesting results in goldpoint geometry may be found as special cases of results in cross-ratio geometry. We do not follow this direction in this book, but intend to do so later.

Lemma 2: The six possible cross-ratios from AGHB are $\pm\alpha^{\pm1}$ and $\alpha^{\pm2}$.

Proof: The cross-ratio (AG, HB) is:

$$\frac{AH}{AB} : \frac{GH}{GB} = \frac{1/\alpha}{1} \times \frac{1/\alpha}{1/\alpha^3} = \alpha = +\alpha^{+1}.$$

Then (see [6]) the other possible cross-ratio values are $1/\alpha, 1/(1-\alpha), (1-\alpha), (\alpha-1)/\alpha, \alpha/(\alpha-1)$, which yield the values given in the lemma. $\qquad\square$

The next lemma concerns two points placed in AB, which, together with A and B form an harmonic range. The first point is a goldpoint of AB.

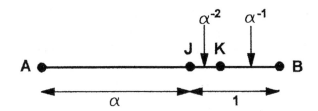

Figure 2. Diagram for Lemma 3

Lemma 3: Let J, K be defined with respect to A, B as follows:

J, K are points internal to the segment AB such that:

$$\frac{AJ}{JB} = 1\alpha, \text{ and } \frac{AK}{KB} = 2\alpha.$$

Then   (i) A, J, K, B is a harmonic range [6];

          (ii) J is a goldpoint of AB;

          (iii) K is a goldpoint of JB;

          (iv) J is a 3rd-order goldpoint of AK.

Proof: We may set $|JB| = 1$, and then the diagram of the range is as shown above, since:

$$
\begin{aligned}
AJ/JB &= \alpha \Rightarrow AJ = \alpha; \\
\text{and} \quad AK/KB &= 2\alpha \\
\Rightarrow \quad AJ + JK &= 2\alpha(JB - JK) \\
\alpha + JK &= 2\alpha - 2\alpha JK \\
\Rightarrow \quad JK &= 1/\alpha^2; \\
\Rightarrow \quad KB &= JB - JK = 1/\alpha; \\
\Rightarrow \quad AK &= \alpha + 1/\alpha^2 = 2, \\
\text{and so} \quad AB &= 2 + 1/\alpha = \alpha^2.
\end{aligned}
$$

Then for (i), the cross ratio is:

$$(KA, BJ) = \frac{KB}{KJ} \times \frac{AJ}{AB} = \frac{1/\alpha}{-1/\alpha^2} \times \frac{\alpha}{\alpha^2} = -1,$$

(The other two cross-ratio values are 2 and 1/2.)

So, by definition, the range is harmonic [6].

Results (ii), (iii) and (iv) follow directly from the diagram measures. □

In view of Lemma 3, we could call (J, K) a *golden-harmonic pair* of points with respect to the segment AB.

After the above preamble, definitions and lemmas, we shall now state and discuss briefly the objectives of goldpoint geometry, as we see them.

The general objectives for studies in Goldpoint Geometry are the discovery and analysis of properties of goldpoints in geometric figures, as the goldpoints arise, naturally or by design, in geometric constructions.

The occurrences of the goldpoints are either by construction, or else by direct introduction of such points into the figures. Further constructions may then be made by which additional goldpoints arise, thereby extending the scope and interest of the associated figures and analyses. Many examples of these procedures are given in this Section.

By contrast, we point out that so-called Fibonacci mathematics is often concerned with study of pure mathematical or natural objects and processes in which discoveries of golden-means or Fibonacci numbers, and identities involving them, bring surprise, delight and new insights into the objects and processes. The goldpoints occur naturally, not by introduction. Many books have been written which detail studies of this nature, across sundry domains. The book by Huntley [3] is a good example. The web-page maintained by Knott [4] may be consulted for many more examples.

The next purpose of this Chapter, following the above introduction, is to give four simple examples of studies in Goldpoint Geometry. The first two examples were introduced briefly in a talk to the Fibonacci Association Conference in July, 2000; they were not, however, presented in the paper submitted for the Conference, so are new* in this book.

The first one concerns the construction of a segment of length α, within the hypotenuse of a $(90°, 60°, 30°)$ triangle. The second example shows the

---

*It is appropriate to acknowledge here that these two examples may be re-discoveries, such is their elementary nature and the antiquity of the subject. However, their presentation in the context of Goldpoint Geometry is surely new.

construction of a snowflake fractal, all of whose points are goldpoints. This fractal will be used and studied in much more detail in Chapter 3.

The third example shows how attractive results about goldpoints of various orders are found when points are introduced into the sides of a square. Squares with goldpoints are treated in more detail in Chapters 4, 5 and 6.

The fourth and final example presents an algorithm for, and solution of, a counting problem which was proposed for a goldpoint set which arises in the formation of a Fibonacci Golden Star [1].

1.2   Constructing a Segment of Length alpha

The construction of a segment of length α, as shown below, is carried out entirely within the well–known $(90°, 60°, 30°)$ triangle of Mechanics problems fame. It also uses the even more famous $(90°, 45°, 45°)$ triangle, whose hypotenuse wrought havoc to the religious beliefs of the Pythagorean School, some 2500 years ago. We shall use the usual triples of sides to denote these two triangles thus: $T_1 \equiv (1, 2, \sqrt{3})$ and $T_2 \equiv (1, 1, \sqrt{2})$. The algorithm to construct the segment can be described in three lines, and proved in one or two, as we shall see. It's description is:

> *Embed T_2 in T_1; 'switch' the hypotenuse of T_2 to span from the common right-angle onto the diagonal of T_1. A segment of length α then appears on the latter diagonal!*

The entire construction is carried out with straight-edge and compass, as shown in Figure 3 below. A final swing of the compass creates a goldpoint in a segment of unit length, if such is desired.

The usual method given for constructing a goldpoint in a unit segment is described in [3] and [4], as it is in all similar texts.

The elegance of our 'new' construction lies in its brevity and wit (if a hypotenuse switch can be so described), and the fact that everything occurs within or between two famous right-angled triangles.

1.21 Construction algorithm for α

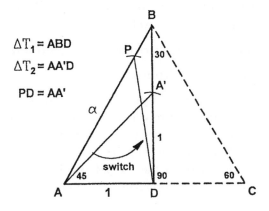

Figure 3. Construction of an α–segment

(1) Construct triangle $T_1(1, 2, \sqrt{3})$ as half an equilateral triangle of side 2. Label it ABD.

(2) Use the compass, with centre on D, radius DA, to mark A' on DB. This completes $AA'D$ as triangle $T_2(1, 1, \sqrt{2})$.

(3) With compass again on centre D, and radius AA', mark off point P on AB (this is the *hypotenuse switch* referred to above).

Theorem 1: $AP = \alpha = (1 + \sqrt{5})/2$

Proof: Let $x = AP$, and consider triangle APD. By the cosine rule, and since $PD = \sqrt{2}$, $x^2 = 2 - 1 + 2.1.x.\cos60° = 1 + x$. Hence $x^2 - x - 1 = 0$. Clearly $x > 1$, hence $x = \alpha$. $\square$

Let us denote the final triangle APD by $T_3(1, \alpha, \sqrt{2})$. A number of interesting consequences follow easily.

Theorem 2:

(i) The new triangle T_3 lies between T_1 and T_2 in area, with $T_2 < T_3 < T_1$.

(ii) Their areas are in proportions $1 : \sqrt{3}\alpha/2 : \sqrt{3}$.

(iii) $AP \times PB = \alpha^{-1}$. Hence P is a 3rd-order goldpoint of AB.

Proof: (i) and (ii) both follow from the fact that all three triangles are on the same base $AD = 1$, and have heights $1, \alpha\sin 60°$ and $\sqrt{3}$ respectively.

(iii) follows from the fact that $PB = 2 - \alpha = \alpha^{-2}$. □

1.22 Further constructions within $T_1(1, 2, \sqrt{3})$

If we use the compasses to mark off just four more points in T_1, we can construct several more segments involving powers of α. Now that the basic idea has been explained with reference to Figure 3, we shall give a second figure which contains all six constructed points; and then we shall list what has been achieved with regard to the six points. Proofs are all simple calculations from triangle and α formulae, and so only hints at them are given.

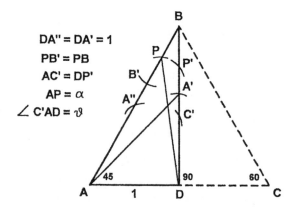

Figure 4. Construction of points and segments in T_1

The constructions will be clear from the dotted lines and small arcs. The points are produced in the following order: A', A'', P (the hypotenuse switch), B', P'.

The following constructions have been achieved:

$$
\begin{aligned}
DA' &= 1 && = DA'' \\
AP &= \alpha && = AP', \text{ (proved above)} \\
A''P &= \alpha^{-1} && (AA'' = 1, \text{ since } ADA'' \text{ is equilateral}) \\
PB &= \alpha^{-2} && (= 2 - \alpha) \\
A''B' &= \alpha^{-3} && (= \alpha^{-1} - \alpha^{-2}) \\
DP' &= \alpha^{1/2} && (= \sqrt{(\alpha^2 - 1)}) \\
\sin\theta &= \alpha^{-1/2} && \text{(see below, for segment } C'D) \\
P'C' &= \alpha^{-3/2} && (= \alpha^{1/2} - \alpha^{-1/2}).
\end{aligned}
$$

We can obtain a segment of length $\alpha^{-1/2}$ by setting the compass at $DP' = \sqrt{\alpha}$ and marking a hypotenuse of that length from A to a new point C' on BD, forming another right-triangle ADC'. Then the required segment is $C'D = \sqrt{(\alpha - 1)} = \alpha^{-1/2}$.

By the above we have achieved, and differently from normal textbook presentations [e.g. [3]], the construction of goldpoints in unit segments: for both $A''B$ and $A''A$ are segments of unit length: and the point P is an interior respectively exterior golden-mean of these two segments.

We note that we could easily have constructed a goldpoint of the original unit segment AD, by placing compass on A, radius AP, and marking P'' on AD produced. Then with compass on D and radius DP'', mark point Q between A and D, and that will be the required goldpoint.

To summarize, beginning with the famous mechanics triangle, and making just four or five swings of the compass, we have constructed within the triangle a large number of goldpoints and powers of α, and observed relationships between three fundamental triangles.

1.3 The Goldpoint Snowflake Fractal

In this second example of 'Goldpoint Geometry in action' we show how to construct a snowflake fractal, all of whose limit points are goldpoints.

Since a variety of applications of this fractal are described in some detail in Chapter 3, we shall simply define it here (i.e. give its base and motif), give pictures of its motif and fifth phase, and compute its fractal dimension using a formula given in [5].

1.31 The base and motif

Following von Koch (1904), [see [5], page 32], the *goldpoint snowflake fractal* is a Cantor-type fractal formed from the unit line-segment $U = [0, 1]$. This segment U is therefore the *base* of the fractal. It is said to be *phase 0* of the recurrence process which converges to the fractal itself.

The segment AB of length 1 (not shown in Fig. 5) is the base, *phase 0*, of the snowflake. To construct the *motif*, which is called *phase 1* of the process, we mark in the goldpoints of segment $AB = U$, which occur at points $(1/\alpha^2, 0)$ and $(1/\alpha, 0)$. Then we remove the set of points which lie between the goldpoints and replace it by an inverted V, whose arms are each of length $1/\alpha^2$.

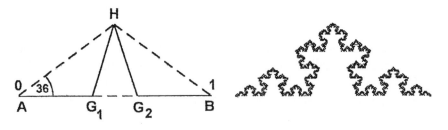

Figure 5(i)   *Figure 5(ii)*
The snowflake motif, phase 1   *The goldpoint snowflake, phase 5*

Figure 5(*i*) shows this motif; and Figure 5(*ii*) shows the goldpoint snowflake after the above process has been re-applied four times, using reduction factor α^2 each time, to all line-segments. The process is to be continued ad infinitum; because of the reduction factor (less than 1), the total length of all segments at a phase will tend to a finite limit.

1.32 Calculation of the snowflake's fractal dimension

The number of segments achieved at each phase is 4 times the number of segments at the previous phase; we denote this multiplication factor by m. And the reduction factor is denoted by r. Then the fractal dimension of the snowflake is:

$$d = \frac{\log m}{\log r} = \frac{\log 4}{\log \alpha^2} = 1.44042\ldots$$

1.4 On Triangles Involving the Golden Mean

The bounding triangle of the snowflake motif (see Figure 5(i)) consists of two triangles similar to AG_1H, and the triangle G_1G_2H. It is easy to calculate the angles and sides of these triangles: their angles are, respectively, $108°, 36°, 36°$ and $72°, 72°, 36°$. Their sides are in ratios indicated in the following designations.

Designating the triangles by S and T respectively, we can write them thus:

$$S \equiv S(1, \alpha^{-1}, \alpha^{-1}) \text{ and } T \equiv T(1, \alpha^{-1/2}, \alpha^{-1/2}) .$$

These triangles are indeed well-known, for the motif appears five times in the construction of the pentagram star within a regular pentagon, a mystical symbol of the Pythagorean Brotherhood and other religions of antiquity. In [3] the second triangle T is called the *Golden Triangle*, whereas in [4] the author refers to them respectively as the 'flat' and the 'sharp' triangle, the former having an obtuse angle and the latter an acute angle at the apex. We shall meet these triangles again, in Chapter 3, when fractals are treated in more detail.

The occurrence of these triangles, having sides of integer or α-power lengths, as well as integer-degree angles, prompts us to propose the following problem:

> Can the class of all triangles which have one side of integer length, at least one angle of integer-degree, and two sides of α-power length (or, more generally, two sides whose lengths are real multiples of α-powers), be characterized?

We have discovered a list of examples[†] in this class (not presented here); but we have no idea how to proceed with the characterization problem.

More generally, we could study triangles whose sides have lengths each of which is a quadratic integer of the field $Q(\sqrt{5})$, that is, each of which are members of $Z(\alpha)$ (see [2] for details of this field).

[†]A right-angled triangle in the class is $(1, 1/\sqrt{\alpha}, \sqrt{\alpha})$

1.5   On a Square with Goldpoints

This example of goldpoint geometry begins with a square, and marks two points on each of two adjacent edges. Further constructions are made, which determine two other points; the somewhat surprising conclusion is that these two are goldpoints with respect to a constructed segment if the first pairs of marks are first- or second-order goldpoints within the edges of the square.

Theorem: Let $ABCD$ be a square.

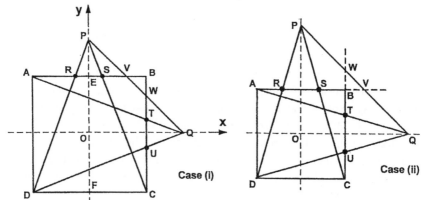

Figure 6. Diagrams for the two cases

Construction: In the side AB mark two distinct points R, S, symmetrically placed with respect to AB. Similarly, in the side BC mark two distinct points T, U symmetrically placed with respect to BC.

Draw DR and CS, and produce them to meet at P. Draw AT and DU, and produce them to meet at Q. Join P to Q. Let the line PQ cut the lines AB, BC in points V, W respectively.

Propositions:

Case (i): If R, S and T, U are goldpoints of AB, BC respectively, then V, W are goldpoints of segment PQ. (PQ cuts the square internally; and V is a third-order goldpoint of AB.)

Case (ii): If R, S and T, U are second-order goldpoints of AB, BC respectively (e.g. $AS/SB = \alpha^2$), then again V, W are goldpoints of segment PQ. (Now PQ 'cuts' the square externally; and V is an external first-order goldpoint of AB.)

Proof of Case (i):

We are concerned with ratios, so we can choose to let the side AB of the square be unity. And we can place Cartesian axes as shown (dotted) with origin at the centre of the square.

Since $|RS| = 1/\alpha^3$, the goldpoint S has coordinates $(1/(2\alpha^3), 1/2)$; and C has coordinates $(1/2, -1/2)$. So the line CS has gradient $(1/2 - (-1/2))/(1/(2\alpha^3) - 1/2) = -\alpha^2$.

Therefore $PF = \alpha^2 FC = (1/2)\alpha^2$.
So $PE = PF - EF = (1/2)\alpha^2 - 1 = 1/(2\alpha)$.
Now, by symmetry, PQ has gradient $-45°$, hence $PV = \sqrt{2}PE = 1/(\sqrt{2}\alpha)$, and $PW = \sqrt{2}(1/2) = 1/\sqrt{2} = VQ$.
Hence $PV/VQ = 1/\alpha$, so V is a goldpoint of PQ.
Similarly, W is the other goldpoint of PQ.

The length of PQ is $PV + PW = (1/\alpha + 1)/\sqrt{2} = \alpha/\sqrt{2}$.

It is evident that V is internal to AB ($EV = PE = 1/(2\alpha) < 1/2$). And since $AV/VB = (1/2 + 1/(2\alpha))/(1/2 - 1/(2\alpha)) = \alpha^3$, we find that V is a third-order goldpoint of AB. $\quad\square$

Proof of Case (ii):

Similar Cartesian analysis establishes the claims of Case (ii).

Corollary to Case (i): Produce AB and DU to meet at K. Let the intersection of DU and CP be J. Then:
$BK = \alpha$ (from $AK/AD = BK/BU$); and the circle drawn on SK as diameter passes through J, since $\angle SJK = 90°$. [N.B. we shall see in Chapter 2 that this circle is the B-ring, or goldpoint ring, of AB. Note that K is an exterior goldpoint of AB.]

The following figure shows what is achieved by carrying out the constructions used for Case (i) for *all* sides of the square $ABCD$.

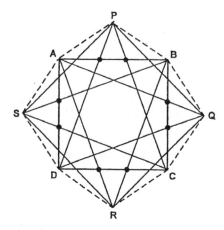

Figure 7. Square and Star constructed from a SGP

It is evident that $PQRS$ is a square; and all its goldpoints are constructed, being the points where its sides cut the sides of $ABCD$.

The interior of the diagram displays an attractive star-like figure, with eight vertices. It is not fully regular, since $ABCD$ and $PQRS$ are not equivalent squares. Some measurements from the diagram, given that $|AB| = 1$, are: $PQ = \sqrt{2}.PO = \sqrt{2}.(\alpha/2) = \alpha/\sqrt{2}$: hence the square $PQRS$ has area $\alpha^2/2$, and diagonal α.

Further developments: We could apply the case (i) constructions directly to the new square $PQRS$ and its goldpoints, and obtain a third square. This would be rotated through $45°$ anti-clockwise from $PQRS$, and hence would be in similar position, and concentric with, the original square $ABCD$. Its area would be $\alpha^4/4$

Evidently we could repeat this process again and again, indefinitely. The result would be a sequence of squares, all with centre O and alternately having sides parallel to $ABCD$ or $PQRS$. Their areas would form the sequence: 1, $\alpha^2/2$, $\alpha^4/4$, $\alpha^6/8$, $\cdots$

A moment's thought, too, reveals a method for constructing a sequence of squares with goldpoints which starts with $ABCD$ and whose members also rotate about O but which shrink in size from one to the next.

Another result about this Star is the following: Diagonals SB and AQ

meet in a point F, say. Then lines DH and CG also meet in F (here we are using G and H to denote the goldpoints of AB). Moreover, the vertical from centre O to P also passes through F; let this vertical bisect AB in point E; then points E and F are the goldpoints of OP.

Many more things could be said about this Star — indeed, there is an embarrassment of goldpoint riches in it — but we must leave it there.

1.6 Fibonacci Golden Stars: Goldpoint Counting

The purpose of this subsection is to show how the points of Star figures may be enumerated and checked for goldponts.

We may write $G^{(\lambda)}$ to denote a goldpoint of multiplicity λ, if it is a goldpoint with respect to λ different[‡] line-segments in the geometric figure. We call λ the *goldpoint number* of the point. The total of goldpoint numbers in a geometric figure may be computed, and a goldpoint density computed from it.

In fact, in this section we treat only the *simple goldpoint count* and *simple goldpoint density*, whereby a point in the figure is counted once if it is a goldpoint in *at least one* line-segment.

We study only one example, namely that of the Fibonacci Golden Star, first introduced by Turner on the 1996 Christmas card described in subsection 1.1 above.

Since the total number of points (say N) on the Fibonacci Star is large, it is difficult even to count them, let alone to decide which of them are goldpoints, and what their goldpoint multiplicities are. It was deemed necessary to develop a few techniques for simplifying the counting and α-analysis procedures. These techniques are to be described next.

We tried several approaches, none of which provided a single algorithm for tackling any given geometric figure. The following notes and methods, however, do give some general insights into these counting problems. For the Fibonacci Star itself, we finally discovered a simple solution method, using similar triangles and Euclidean geometry, which completes most of

[‡]Define $G^{(0)}$ to be a null-goldpoint (that is, a point which is not a goldpoint of any segment).

the analysis; we present that (in chronological order!) as a postscript to
this Section.

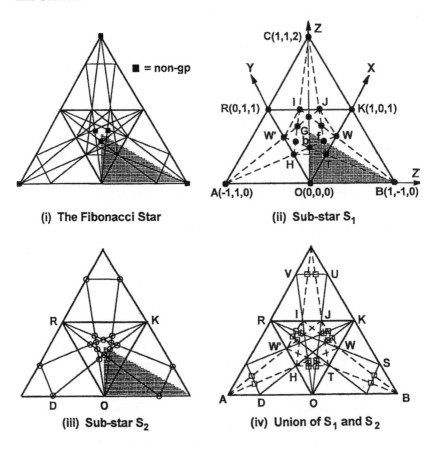

Figure 8. The Fibonacci Golden Star, and two sub-Stars

Using symmetries to partition the figure

In Figure 8 above, we show first the Fibonacci Star from the Christmas
card, and with it two less complicated Stars (or sub-Stars), labelled S_1 and
S_2 respectively. It is easy to see that the union of these two graphs equals
the first one, i.e. $S = S_1 \cup S_2$. Note that we use the symbol $\cup$ to mean an
operation of superposition and graph union: the two sub-stars are drawn
to the same scale. This union graph is also shown.

Counting crossing-points in the Stars

In the Fibonacci star, diagram 8(i), a lower-right, right-angled triangle is shown shaded. This triangle is formed under the intersection and between two medians. There are six triangles similar to this one, all having the same geometric design, whose union comprises the whole Fibonacci Star. In this union of six right-angled triangles, the points on the hypotenuse of a triangle 'occur' three times, and so do the points on the short side. Whereas, the points in the interior and on the third side 'occur' six times.

Using these facts, we can easily count the number of points in each of the three stars shown above. The results are in the following table.

Number of points	Star S_1	Star S_2	Fibonacci Star
Interior and third side	1×6	3×6	6×6
hypotenuse and short side	4×3	4×3	6×3
Totals	18	30	54

There appear to be discrepancies in the totals row of the table; but it must be observed that (a) points counted in S_2 were already counted in S_1, and (b) when the union is formed in fig. (iv) new points arise at intersections of full and dotted lines, as shown by open squares. In fact, the total of points in the Fibonnaci Star is the sum of full dots, open dots, and open squares in figs. (ii), (iii) and (iv) respectively, giving a total of 54, as in the table.

We now turn to the problem of counting goldpoints in the Stars.

Goldpoints in the sub-star S_1

In diagram 8(ii) we have shown axes OX, OY, OZ, placing the triangle in the honeycomb plane $x + y = z$ as shown; the coordinates of some of the points are given. [N.B. We could have worked in the xy-plane, but for this work we preferred using the honeycomb plane.]

G is the centroid of $\triangle ABC$, which is the union of six triangles equivalent to the shaded one (i.e. $\triangle OBG$). G has coordinates $(1,1,2)/3$.

(i) *The star's points:* In sub-star S_1 all the intersection points are shown with full dots. On the boundary sides BG and GO of $\triangle OBG$ there are the four dots (points) O, b, f, B. There are no dots in the open third side; but in the interior is the one point T. These observations confirm the result tabulated above, namely that this Star has a total number of points:

$$\#S_1 = 4 \times 3 + 1 \times 6 = 18 \,.$$

[N.B. Of course, it is easy to obtain this count directly from the well-drawn figure 8(ii). It is much more difficult to count the points of the Fibonacci Star in figure 8(i) directly.]

(ii) *The Star's goldpoints:* To determine which points in S_1 are gold-points, we only have to examine which of $\{O, b, f, B\}$ and $\{T\}$ are gold-points, and multiply their numbers by 3 and 6 respectively.

Now O and B are clearly not goldpoints of any line-segment in S_1; whereas T is a goldpoint, of both OW and OK, by the construction of the Fibonacci Star and similar triangles. Hence only b and f remain to be checked. Their coordinates are easily found to be $b = \alpha^{-3}(1, 1, 2)$ and $f = [1/(2\alpha - 1)(1, \alpha^{-3}, 2\alpha)]$, using equations of line-pairs (BW', AW) and (CT, AW) respectively. [Incidental observation: b is a 3rd-order goldpoint of the median OC.]

Since the arrangements of points on line-segments AW, BW' and CT are identical, we only need to study the points on one of them: we chose AW, and the following diagram shows all the coordinates of the points on it.

A	H	b	f	W
(-1,1,0)	$\tau^2(0,1,1)$	$\tau^3(1,1,2)$	as above	$\tau(1,0,1)$

For b, we have to check $[A, b, W], [a, b, f], [H, b, f], [H, b, W]$; whereas for f we must check $[A, f, W], [H, f, W], [b, f, W]$.

It is often found that once two or three such segments have been checked, to see whether the inner point is a goldpoint, other segments can be discarded from consideration. Nevertheless, this work is very tedious to carry out.

We found that neither b nor f are goldpoints in S_1. Hence, finally for S_1, there are 18 points, of which 6 (of type T) are goldpoints. The simple goldpoint density of S_1 is therefore $6/18 = 1/3$.

Goldpoints in the sub-Star S_2

We can use the same methods as above to determine the goldpoint density of S_2. However, we remark that with hindsight, after drawing the diagrams of Figure 8 using CAD software on a PC, we realised that we could place the figure (iii) on a 1cm-scale grid on the PC screen, and use the mouse to discover directly the (approximate) coordinates of the points on line-segment DK. This was all we needed for discovering candidates for goldpoints. Designating the interior crossing-points of DK by H, p, q, r, s, the following diagram results (N.B. t and u are further points, occurring in the union fig. (iv)):

	D		H	p	q	r	s	(t	u)	K
x:	-2.54		-.81	-.27	0	.32	.62			2
y:	0		1.31	1.75	1.94	2.21	2.42			3.51

The point p lies in four segments, namely Dq, Dr, Ds, DK. Given the measurements of the above diagram, it is quickly established that p is not a goldpoint in any of the four segments. Similarly for q, in its six segments. However, it is found that both r and s appear to be goldpoints, in segments DK and pK respectively: the ratios concerned, namely $Dr : rK$ and $sK : ps$, are calculated to be approximately 1.65. We check them geometrically, as follows:

The point r is the intersection of lines DK and OU (see fig. (iv)). Now U is a goldpoint of CK and is found to have coordinates $(1, \tau, \alpha)$, where $\tau = 1/\alpha$. Hence the lines are:

$$
\begin{aligned}
DK &\equiv & -\tau(x-1) &= & y/\tau &= & 1-z \quad (= s, \text{ say}) \\
OU &\equiv & x-1 &= & \alpha(y-\tau) &= & \tau(z-\alpha) \quad (= t)
\end{aligned}
$$

From these line equations we calculate the coordinates of r to be $\tau(\tau, \tau^2, 1)$.

Next we calculate the corresponding goldpoint of DK thus:

$$
\begin{aligned}
\text{goldpoint} &= \tau^2(1\underline{D} + \alpha\underline{K}) \\
&= \tau^2[(-\tau, \tau, 0) + \alpha(1, 0, 1)] = r \,.
\end{aligned}
$$

Thus we have shown that indeed r is a goldpoint in DK. Similarly we may confirm that s is a goldpoint in pK.

Finally we observe that in all the six line-segments which correspond to DK in the sub-star S_2, nine of the twelve central points (marked with an open circle in fig. (iii)) correspond to either r or s and hence are themselves goldpoints. The other three, marked with a black square in fig. (i), are not goldpoints of any segment. We also observe that all of the nine points on the triangle ΔOKR are goldpoints, as previously demonstrated.

Hence the total count of goldpoints in S_2 is $9 + 9 + 6 = 24$. Dividing by the total number of points gives the simple goldpoint density of S_2 to be $24/30 = 4/5 = 80\%$, a pleasingly high result.

Goldpoints in the Fibonacci Star

Finally, we address the Christmas puzzle which introduced this chapter. [In retrospect, we should say that this is not suitable puzzle-fare for Christmas day! Indeed, one might use up all the twelve days of Christmas in search of its solution — unless one hits upon the method which is shown at the end and labelled Postscript!]

We have already discovered that the Fibonacci Star has 54 points. And we can count up the goldpoints found in sub-stars S_1 and S_2. But now we must take into account the new points formed in the union of S_1 and S_2. We must place these points into the line-segments DK and AW, and check them off for new goldpoints: furthermore, we must re-examine the points which were found *not* to be goldpoints earlier, in the light of the newly introduced points.

After doing all that, we find that, finally, the only non-goldpoints in the Fibonacci Star are the six shown by black squares in fig. (i) above. These are the three outer-corner vertices, and the three innermost points. This gives a total of 48 goldpoints in the figure, out of the 54 points in the Star. Thus the simple goldpoint density is $48/54 = 8/9 \approx 89\%$ — a very high density indeed.

Postscript!

The whole process described above was lengthy and extremely tedious, relieved only somewhat by the attractiveness of the diagrams, and the pleasing final result. However, it can be much curtailed, using similar triangles,[§] as follows:

[§]The author, to his simultaneous shame and joy, only spotted these triangles after the above treatment of the problem was typeset and the diagrams were being drawn using a CAD package. He decided to leave the preliminary, lengthy analysis herein, since it demonstrates many interesting points about the Fibonacci Star and its two sub-Stars.

(1) H is a goldpoint in OW'. It follows by similar triangles in $\triangle OJA$, with $DH\|AJ$, that p and r are goldpoints.

(2) s is a goldpoint since it corresponds in segment RS to point p in segment DK.

(3) t is a goldpoint in $\triangle RTK$, since $IW\|RT$, and I, W are goldpoints in RK and TK respectively.

(4) u is a goldpoint in HK, since $Hu/uK = RJ/JK$.

(5) The points on AW can be dealt with similarly, using $\triangle OKW'$, with $HW\|W'K$.

(6) The quite major task that remains is to show that q is **not** a goldpoint. This involves checking a ratio in each of the fifteen segments along DK which include q. Many can be quickly dismissed, but it seems that coordinate geometry has still to be used, for at least part of this task.

Chapter 2

The Goldpoint Rings of a
Line-Segment

2.1 Definition of Goldpoint Rings

In this chapter we introduce the notion of *goldpoint rings with respect to a line segment*, which is a generalization of the concept of goldpoints in a line segment. We begin with a definition of goldpoint rings, and then study some of their properties.

> **Definition 2.1:** Let AB be a line segment in a plane, and P be a point in the plane which satisfies one of the following two conditions: (i) $AP/PB = \alpha$; (ii) $AP/PB = 1/\alpha$. Then the two loci of P determined by the conditions are called *the goldpoint rings of AB.*

We shall see below that one ring is a circle that contains point A (we shall call this the *A-ring of AB*); and the other is a circle containing point B (which we shall call the *B-ring of AB*). Sometimes we shall refer to them as the α-rings of AB.

Note that the goldpoint rings contain the goldpoints (both interior and exterior) of AB. These are, of course, where the loci of P cuts the line of AB produced in both directions.

Our first task is to show that the name 'ring' is appropriate for these loci. We shall do this by showing that the rings are in fact two circles, placed symetrically with respect to the ends of AB. The arrangement of AB and its two goldpoint rings in the plane has a diagram which looks much like a pair of spectacles. The centre of the A-ring is on BA produced, whereas the centre of the B-ring is on AB produced. Later we shall compute formulae for the positions of these centres, and show that each ring has radius $|AB|$.

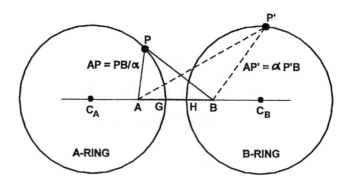

Figure 1. Segment AB with its goldpoint rings

Note that G and H are the goldpoints of AB. The radius of each gold-point ring is $|AB|$, as we shall prove shortly.

2.2 The General ρ-rings Relative to AB

It will be convenient to solve a more general problem first, taking $AP/PB = \rho$ with $0 < \rho < \infty$; that is, using a general ratio instead of the golden mean α or its reciprocal.

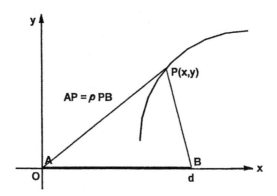

Figure 2. Diagram with $AB = [0, d]$, and $AP/PB = \rho$

Without much loss of generality, we can place AB and P in the Cartesian xy-plane, and let AB be the segment $[0, d]$. Then we can announce the details of several rings with specially chosen values of ρ. Figure 2 is a general diagram for the situation.

The locus of P is obtained thus:

$$
\begin{aligned}
AP^2 &= PB^2\rho^2 \\
x^2 + y^2 &= (x - d)^2\rho^2 + y^2\rho^2 \\
x^2(1 - \rho^2) + y^2(1 - \rho^2) &= -2dx + d^2\rho^2 \\
x^2 + y^2 + 2dx\rho^2/(1 - \rho^2) &= \rho^2/(1 - \rho^2) \ (\rho \neq 1).
\end{aligned}
$$

Writing $u = \rho^2/(1 - \rho^2)$, and completing the square for x we get:

$$
(x + du)^2 + y^2 = d^2u(u + 1).
$$

Thus the locus of P is a circle. It has centre $C(-du, 0)$ and radius $d\sqrt{u(u + 1)}$. Each value of ρ $(\neq 1)$ determines a circle. We shall call these circles ρ_{AB}-rings, and note the following theorem about them.

Theorem 1: Given any segment AB of length d, then the set of its ρ_{AB}-rings, together with the vertical line $x = d/2$ (which is the '$\rho = 1$ ring of infinite radius') partitions the set of points in the xy-plane.

Proof: Any finite point $P(x, y)$ in the plane determines a unique value of $\rho(= AP/PB)$ which determines the ρ-ring to which P belongs. And no two ρ_{AB}-rings intersect (finitely), otherwise an intersection point would have two values of ρ, which is impossible. □

Definition: The two rings for which $\rho = r$ and $\rho = 1/r$ are *complementary ρ_{AB}-rings*.

The circles for a pair of complementary rings have u-values of $r^2/(1-r^2)$ and $1/(r^2-1)$ respectively. Using the formulae given above, these determine their circle equations, centres and radii directly.
They are of equal radius, $R = |r/(1-r^2)|$, since $u(u+1) = r^2/(1-r^2)^2$ in each case. Their common area is therefore $\pi d^2 r^2/(1 - r^2)^2$.

We are specially interested in the goldpoint rings, and also other cases where ρ is a power of α. A table of the simpler cases follows. We give details of the 'infinite ring' (the line $x = 1/2$) , and three pairs of complementary

ρ-rings, taking AB to be the interval $[0, 1]$.

ρ	1	$\alpha^{1/2}$	$\alpha^{-1/2}$	α	α^{-1}	$\alpha^{3/2}$	$\alpha^{-3/2}$
$-u(centre)$	∞	α^2	$-\alpha$	α	$-\alpha^{-1}$	$\alpha^2/2$	$-\alpha^{-1}/2$
$R(radius)$	∞	$\alpha^{3/2}$	$\alpha^{3/2}$	1	1	$\alpha^{1/2}/2$	$\alpha^{1/2}/2$

A sketch of these ρ-rings indicates how the B-rings are nested to the right of the line $x = 1/2$, with their centres approaching $B(1, 0)$ from above as ρ decreases to one. The A-rings, with $\rho < 1$, are mirror images (in $x = 1/2$) of the B-rings. The whole picture is somewhat reminiscent of a strangely oriented magnetic field in a plane, with AB being the magnet.

2.3　The B-ring of AB, with $AB = [0, 1]$

We now study some properties of the B-ring ($\rho = \alpha$) of a segment AB of unit length. The following figure is used for the next two theorems.

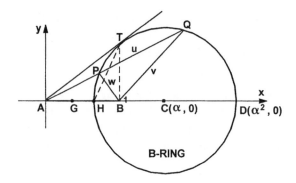

Figure 3. Segment AB, with its B-ring

Constructions:
　　$AB = [0, 1]$, and G, H are its interior goldpoints.
　　The B-ring has centre $C(\alpha, 0)$, radius 1; and it cuts the
　　x-axis at the goldpoints $H(1/\alpha, 0)$ and $D(\alpha^2, 0)$ of AB.
　　APQ is any ray from A which cuts the B-ring in P, Q.
　　AT is a tangent to the B-ring. $PQ = u$, $QB = v$, $BP = w$.

Theorem 2:

$$
\begin{aligned}
(i) \quad && AT &= \sqrt{\alpha} \\
(ii) \quad && TB &= \alpha^{-1/2} \text{ and } TB \perp AB \\
(iii) \quad && HT &= 2/\alpha \\
(iv) \quad && wv &= 1/\alpha \\
(v) \quad u/(v-w) &= & \alpha \,, \text{ hence } uvw = v - w.
\end{aligned}
$$

Proof:

(i) $AT^2 = AH.AD = \alpha \Rightarrow AT = \sqrt{\alpha}$;

(ii) $AT/TB = \alpha$ therefore $TB = AT/\alpha = \alpha^{-1/2}$;
The converse of Pythagoras' theorem on $\triangle ABT$
shows that $TB \perp AB$;

(iii) $HT^2 = HB^2 + BT^2 = \alpha^{-4} + \alpha^{-1} = 4\alpha^{-2}$;

(iv) and (v) $AP = w\alpha$ and $AP = u = v\alpha$ since P and Q are
on the B-ring. Also $AP(AP + u) = AT^2 = \alpha$.
Therefore $w\alpha.v\alpha = \alpha \Rightarrow wv = 1/\alpha$ and $u = (v - w)\alpha$. Elim-
inating α gives $uvw = v - w$. $\qquad \square$

Theorem 3:

If $\triangle PQB$ is isosceles, then P is a goldpoint of AQ.

Proof:
$\triangle PQB$ is isosceles if either $u = v$, or $v = w$, or $w = u$. The
case $v = w$ is impossible, in view of Thm. 2(v) above.

Case $u = v$: By Thm. 2(v), $v = \alpha(v - w) \Rightarrow v = w\alpha^2 = u$. Now $AP = w\alpha$, since P is on the B-ring. Therefore
$AP/PQ = w\alpha/u = w\alpha/w\alpha^2 = 1/\alpha$. Hence P is a goldpoint
of AQ. (Note that $\triangle PQB$ is now $(u, v, w) = \alpha^{1/2}(1, 1, \alpha^{-2})$).

Case $w = u$: $AP/PQ = w\alpha/u = w\alpha/w = \alpha$, hence P is
a goldpoint of AQ. (Note that $\triangle PQB = \alpha^{-2}(1, \alpha, 1)$; it is
similar to the 'sharp' golden triangle.) $\qquad \square$

2.4   A Sequence of Goldpoint Rings

The following equation determines (with parameter i) an interesting sequence of goldpoint rings. The case $i = 1$ is the one discussed in subsection 2.3.

$$(x - \alpha^i)^2 + y^2 = \alpha^{2(i-1)} \ .$$

We shall first give a table for four rings in the sequence, showing their main features. Figure 4 below shows how they relate to one another. Some construction lines are added, such as their common tangent lines from A. After the figure we list some of their interesting properties as a theorem.

i	rel. segment	Centre C_i	Radius R_i
-2	$AI = [0, 1/\alpha^3]$	$H(1/\alpha^2, 0)$	$1/\alpha^3$
-1	$AH = [0, 1/\alpha^2]$	$G(1/\alpha, 0)$	$1/\alpha^2$
0	$AG = [0, 1/\alpha]$	$B(1, 0)$	$1/\alpha$
1	$AB = [0, 1]$	$C(\alpha, 0)$	1

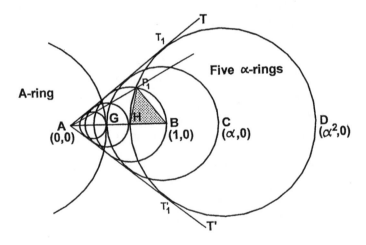

Figure 4. Five α-rings in a general sequence

It is evident from the table and figure, that each relative segment is a golden section segment of the previous one. And that this sequence of rings could be extended indefinitely in either direction, as i ranges over $\mathbf{Z}$.

The following theorem lists five properties of this sequence of goldpoint rings. The proofs follow directly from earlier observations and theorems, or by elementary geometry.

Theorem 4:

(i) The α-rings in the sequence are enveloped by the two lines $y = \pm\alpha^{-1/2}x$, which are tangential to all the circles.

(ii) The i-ring (relative to segment S_i say) touches the $(i-3)$-ring in a goldpoint of S_i. And the $(i-1)$-ring touches the $(i-4)$-ring in the other goldpoint of S_i.

(iii) The $(i+1)$-ring touches the complementary ring of the i-ring in a goldpoint of S_i.

(iv) Let the i-ring meet the $(i-2)$-ring in point P_i. Then the triangle under P_i (see shaded example) is similar to $(1, 1, \alpha^{1/2})$.
And the locus of P_i is a line through A.
The triangle AP_iC_{i-2} is similar to $(\alpha^{1/2}, \alpha^{-1/2}, \alpha)$.

(v) The $(i-2)$-ring is inscribed in $\triangle AT_iT_i'$.

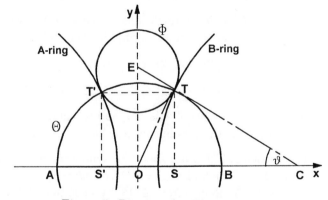

Figure 5. Diagram for Theorem 5

Theorem 5:

(i) The circle Θ, drawn on AB as diameter, is orthogonal to the α_{AB}-rings.

(ii) Let $AB = [-1/2, 1/2]$: and let the circle Θ intersect the A-ring and B-ring in T' and T respectively (upper half of plane). Then the circle Φ which touches the α_{AB}-rings at T' and T has radius $1/4$ and centre $E(0, \sqrt{5}/4)$.

(iii) Let the abscissae of T' and T be respectively S' and S. Then $S'STT'$ is a square.

Proof:

(i) Consider the triangle $\triangle OTC$ (see Fig. 5). We know that $OT = 1/2$, by construction, and that $CT = 1$, the radius of the B-ring: and that $OC = OB + BC = 1/2 + 1/\alpha = (1/2)(2\alpha - 1)$. Therefore:

$$OC^2 = (1/4)(2\alpha - 1)^2 = 5/4 = OT^2 + CT^2.$$

By the converse of Pythagoras' theorem, $\angle OTC = \pi/2$. Hence OT is tangent to the B-ring at T, and CT is tangent to the circle Θ: so the B-ring and Θ intersect orthogonally. Similarly, the A-ring and Θ intersect orthogonally at T'. It follows that Θ is orthogonal to the α_{AB}-rings.

(ii) By (i) CTE is a straight line, of length $r + 1$, where r is the radius of Φ. By similar triangles inside $\triangle EOC$,

$$r/(1/2) = (1/2)/1 \Rightarrow r = 1/4.$$

From $\triangle OTE$, we find that $OE^2 = 1/16 + 1/4 = 5/16$.

Therefore the radius of Φ is $1/4$, and its centre is $(0, \sqrt{5}/4)$. [Note that the three radii concerned are $1/4$, $1/2$ and 1.]

(iii) Evidently $T'T$ is horizontal, by symmetry; and $TS \| T'S'$. From $\triangle OTC$ we find $\tan\theta = 1/2$; so $\sin\theta = 1/\sqrt{5}$ and $\cos\theta = 2/\sqrt{5}$. Hence, from $\triangle OST$ we compute $ST = 1/\sqrt{5}$ and $OS = 1/(2\sqrt{5})$. Therefore $S'S = 2OS = ST$, and so $S'STT'$ is a square. $\qquad\square$

Theorem 6: Given segment AB, having A-ring and B-ring with centres C' and C respectively. Then the ellipse having major diameter CC' and minor radius of length AB has foci A and B. The eccentricity $e = AB/CC'$.

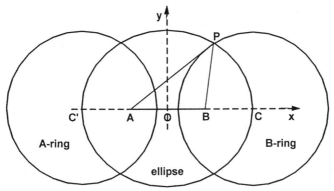

Figure 6. Diagram for theorem 6.

Proof: We shall take $|AB| = 1$. P is an intersection of ellipse and B-ring. Then $C'C = AB + 2BC = 1 + 2/\alpha = (\alpha + 2)/\alpha$. The eccentricity of the ellipse is given by $e^2 = 1 - b^2/a^2$, with $a = CC'/2$ and $b = |AB| = 1$. Simple algebra gives $e = \alpha/(\alpha + 2)$. Therefore $e = 1/CC'$. If S is the right-hand focus, then $OS = ae = 1/2$. Therefore $S = B$; similarly, the other focus of the ellipse is at A. $\qquad\square$

We find the ellipse to be $4x^2 + 5y^2 = 5$; and $AP.PB = 5/\alpha^3$.

Chapter 3

Some Fractals in Goldpoint Geometry

3.1 Introduction

As we have already seen (in 1.7, for example), problems of interest in gold-point geometry [8; 9] arise from study of figures that are obtained when goldpoints are marked on sides of triangles, squares, pentagons etc. and joined by lines in various ways. In Chapter 4 we shall designate such diagrams as *tile-figures* or *golden tiles*.

Many combinatoric problems arise naturally when studying such objects. Another type of problem is to determine how to combine collections of golden tiles in jig-saw fashion, so that they tile a given geometric figure (or the whole plane) with goldpoint marks on touching sides corresponding everywhere. We treat these types of problem in Chapter 4.

Before dealing with tiling and combinatoric problems, however, we wish to present more geometric figures which involve goldpoints, thereby extending the scope of our enquiries. The new figures to be presented are *fractals*, which are formed as sequences of self-similar figures; in many of our examples we shall find that the end result (limiting situation) of the sequencing process is an infinite set of points all of which are goldpoints (or multiple goldpoints) with respect to other points in the set. In the literature on fractals, the limit sets are known as *dust-sets*. We shall call our figures *goldpoint fractals*; and be tempted to call the limit sets of goldpoint fractals *gold-dust sets*!

We shall study a variety of fractals which are achieved by using as base the segment [0,1], and a motif which involves the goldpoints of that segment.*

---

*The terms 'base' and 'motif' are now well-known. Excellent references for these terms, and for several of the analytic techniques used in this chapter, are [3],[6],[11].

In the first three sections goldpoint fractals are described which are dedicated to the memory of the inspirational American mathematician Herta Freitag, who passed away early in 2000, in her 91st year.

In the final section studies of fractals are presented which are based on the regular pentagon. It is well-known (indeed the knowledge goes back to antiquity, since it is mentioned in caballistic literature) that the golden mean occurs frequently in the geometry of the pentagon [3] and its accompanying pentagram star. It is hoped that the results given below on pentagon goldpoint fractals will add to existing literature on the pentagram.

3.2 The Goldpoint Shield (a Golden Snowflake)

The work of this section originated with a fractal picture, which Turner developed for a card to send to Herta Freitag for her 90th birthday. This occurred in November 1998; and just five months earlier Herta had attended the eighth Biennial Conference of The Fibonacci Mathematics Association, where she presented no fewer than three papers. She had attended all the biennial conferences, the first of which was held in 1984; and with her radiant personality, excellent research and delightful presentations, she was an inspiration to all the attendees.

To obtain her birthday fractal picture, Turner took an equilateral triangle, inserted all six goldpoints on its sides, and produced the first five phases of a snowflake exterior to it. This was by the same construction as for the well-known von Koch (1904) snowflake (or island), except that he discarded the open sets between the goldpoints of edges, rather than their middle-third sets.

Below is shown the starting triangle (phase 0), marked with goldpoints, and phases 1, 2, 3 and 5 of the developing snowflake. Since the snowflake is evidently bounded by a non-regular hexagon, and looks very much like a shield that a gallant knight would hold on his arm, the figure was dubbed *Herta's Shield*. On her birthday card the hope was expressed that the shield would keep her safe for many more years to come. Sadly that did not come to pass.

After the diagrams, formulae are given for the growing snowflake's perimeter, and its fractal dimension is computed.

Formulae: In the subsections below, much use will be made of the following identities which relate negative powers of α (i.e. the golden mean $(1 + \sqrt{5})/2$) to themselves and to the ordinary Fibonacci numbers.

$$
\begin{aligned}
(i) \quad & F_{-n} = (-1)^{n+1} F_n; \\
(ii) \quad & \alpha^{-n} = \alpha^{-(n+1)} + \alpha^{-(n+2)}; \\
(iii) \quad & \alpha^{-n} = F_{-n}\alpha + F_{-(n+1)}.
\end{aligned}
$$

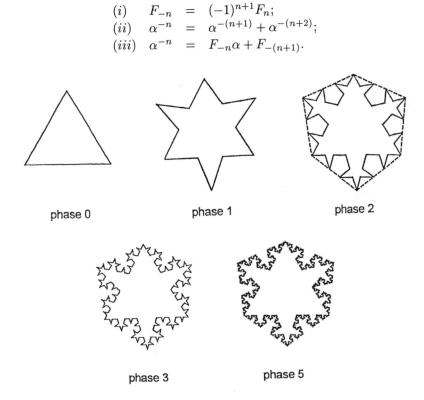

phase 0   phase 1   phase 2

phase 3   phase 5

Figure 1. The goldpoint shield
[The dotted bounding-polygon in phase 2 indicates the shield's outer shape.]

3.21 The fractal dimension of the goldpoint shield

If each side of the starting equilateral triangle (see phase 0 in the figure above), is taken as a *base* of length 1 (unity), then at phase $p = 1$ each side is replaced by the motif which consists of four segments of length $1/\alpha^2$, arranged as shown in the second figure. Since there are 12 segments in all, the snowflake at phase 1 has perimeter length $P_1 = 12/\alpha^2$. Then, at each phase the lengths of segments used are reduced by a factor of α^2, and their total number is increased by a factor of 4. So the formula for the perimeter length at phase p is given by a recurrence as follows:

$$P_p = \frac{4}{\alpha^2} P_{p-1} \ , \ \text{with } P_0 = 3, \ p > 0.$$

Solving the recurrence gives:

$$P_p = 3\left(\frac{4}{\alpha^2}\right)^p \ , \ \text{for } p = 0, 1, 2, \dots$$

The fractal (or self-similarity) dimension (see [3]) is: $d = \log m / \log r$ where $r = \alpha^2$ is the reduction factor, and $m = 4$ is the number of new segments formed at stage p from a segment at stage $p - 1$.

$$\text{Thus } d = \frac{\log 4}{2 \log \alpha} = 1.44042\dots$$

The phase 5 diagram shows how the shield is converging towards a snowflake fractal. Next we examine the goldpoint dust-set, and derive other fractals.

3.3   The Goldpoint Dust-set, and the Comb and Jewel

3.31 The goldpoint dustset

We define the goldpoint dust-set (the *gp-dustset*) by prescribing an infinite process similar to that used to produce Cantor's fractal set.

We take the unit line-segment $[0, 1]$ on the x-axis, and compute its goldpoints, which are at points $(\alpha^{-1}, 0)$ and $(\alpha^{-2}, 0)$; call these points G_1 and G_2 respectively. Then we discard all points in the open set of the segment $(G_1 G_2)$.

Next we compute the positions of the goldpoints of the two remaining segments $[0, G_2]$ and $[G_1, 1]$. Then we discard the two open sets between the two pairs of goldpoints.

We continue this process *ad infinitum*, at each stage discarding all the central open sets between pairs of goldpoints. The limiting set of points (all goldpoints, note), minus the two endpoints $(0, 0)$ and $(0, 1)$, is called the *goldpoint dust-set*.

The computations below Figure 2 describe some properties of this set.

Figure 2. The goldpoint dust-set (to stage 2)

G_1 and G_2 are the goldpoints of line segment $[0, 1]$, and G_3, G_4 are the goldpoints of $[0G_2]$.

Measuring lengths from 0, and writing G_i for $|[0G_i]|$, we find:

$$
\begin{aligned}
G_1 &= 1/\alpha^2 + 1/\alpha^3 = 1/\alpha & &= F_{-1}\alpha + F_{-2} &= \alpha^{-1} \\
G_2 &= 1/\alpha^2 = -\alpha + 2 & &= F_{-2}\alpha + F_{-3} &= \alpha^{-2} \\
G_3 &= 1/\alpha^4 + 1/\alpha^5 = 1/\alpha^3 & &= F_{-3}\alpha + F_{-4} &= \alpha^{-3} \\
G_4 &= 1/\alpha^4 & &= F_{-4}\alpha + F_{-5} &= \alpha^{-4}
\end{aligned}
$$

and so on.

Similarly, H_1, H_2 are the goldpoints of line segment $[G_1, 1]$, and for them we find:

$$
\begin{aligned}
H_1 &= 1/\alpha + 1/\alpha^4 + 1/\alpha^5 &= 1/\alpha + 1/\alpha^3 &= 3\alpha - 4 \\
H_2 &= 1/\alpha + 1/\alpha^4 & &= -2\alpha + 4
\end{aligned}
$$

It may be noted that:

G_1 is a goldpoint of $[0, 1]$ (given)

G_1 is a goldpoint of $[G_2 H_2]$ (since $G_2 G_1 = \alpha^{-3}$ and $G_1 H_2 = \alpha^{-4}$)

G_1 is a goldpoint of $[G_3 H_1]$ (since $G_3 G_1 = \alpha^{-2}$ and $G_1 H_1 = \alpha^{-3}$)

It follows that, as the process of discarding central open segments continues, all of the points left in the dust-set are goldpoints (except 0 and 1): and in the limit, each point is a goldpoint an infinite number of times, with respect to pairs of other points in the dust-set. It might be appropriate to call this the gold-dust set.

It is evident from the above analysis that each goldpoint in the dust-set can be expressed uniquely in α-nary form thus:

$$
\text{goldpoint} = 0.c_1 c_2 c_3 \cdots \equiv c_1 \alpha^{-1} + c_2 \alpha^{-2} + c_3 \alpha^{-3} + \cdots
$$

where all the c_i coefficients are zero or unity, and with no pair of adjacent coefficients being $(1,1)$.[†]

[†]If in the calculation of a goldpoint we obtain both $c_i = 1$ and $c_{i+1} = 1$ we are required to combine the adjacent terms, using $\alpha^{-i} + \alpha^{-(i+1)} = \alpha^{-(i-1)}$.

Examples:

$$G_1 = 0.1, \ G_2 = 0.01, \ G_3 = 0.001, \ \text{etc.}$$

$$H_1 = G_1 + \alpha^{-4} + \alpha^{-5} = G_1 + \alpha^{-3} = 0.101 \, ;$$

$$H_2 = G_1 + \alpha^{-4} = 0.1001 \, .$$

The goldpoint dust-set is the set of all points in $(0, 1)$ which have this type of α-nary form (reminiscent of maximal Zeckendorf representations of n in terms of the Fibonacci numbers).

3.32 The goldpoint comb

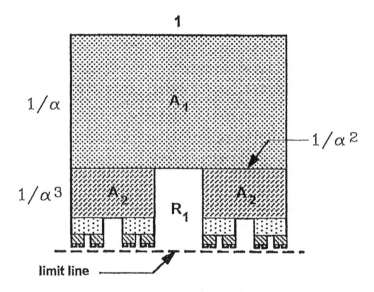

Figure 3. The goldpoint comb

Not long after her 90th birthday, Turner had an email message to say that Herta had had to go into a nursing home. He tried to think of some mathematical item that might make her smile. He wrote to her to say that perhaps she she had not worn the goldpoint shield constantly: hence her recent illness. He included with his letter two further talismans, to add (as he wrote) to her comfort and protection. They were a goldpoint comb for her hair, and a fractal star jewel to ward off evil spirits. He knew something was sadly amiss when he didn't receive a reply to this letter.

The figure above shows Herta's goldpoint comb (imagined to be made of ivory). In the limit it has an infinite number of teeth, the 'prong' points forming a set of measure zero.

It is easy to see how the comb is built up of rectangles, erected upon line segments parallel to those 'left in' during the process of obtaining the goldpoint set (see Fig. 2). Upon each segment a golden rectangle is constructed, with the horizontal segment being the larger side.

Figure 3 shows how the short sides of the rectangles have lengths in the sequence:

$$\frac{1}{\alpha}, \frac{1}{\alpha^3}, \frac{1}{\alpha^5}, \frac{1}{\alpha^7}, \cdots$$

This is a geometric progression of common ratio $1/\alpha^2$, and its infinite sum is 1. Therefore the goldpoint comb has height 1, and it covers (in the limit, and except for the limit line) a square of side 1.

Thus the unit square of the comb is tiled by golden rectangles in a most interesting way.

If we check the 'hole' or 'spaces' in the comb, we see that they are also rectangles, all standing on the horizontal limit line where the teeth 'end'. Again checking the dimensions, we see that each of these rectangles is also a golden rectangle. Moreover, the largest 'hole' rectangle is equal to the second largest ivory rectangle; the second largest 'hole' rectangle is equal to the third largest ivory rectangle; and so on.

3.33 The area (A) of the ivory, and the area (H) of the 'holes'

Working directly from Fig. 3, we get for the total ivory in the comb:

$$
\begin{aligned}
A &= 1 \times \tfrac{1}{\alpha} + 2 \times \tfrac{1}{\alpha^5} + 4 \times \tfrac{1}{\alpha^9} + 8 \times \tfrac{1}{\alpha^{13}} + \cdots \\
&= \sum_{i=1}^{\infty} 2^i (1/\alpha^{4i-3}) \,. \\
&= \sum_{i=1}^{\infty} (2/\alpha)^i (1/\alpha^3)^{i-1} \\
&= \sum_{i=1}^{\infty} (2\alpha - 2)^i (2\alpha - 3)^{i-1} = \alpha^2/3 \,.
\end{aligned}
$$

An alternative way to find A, using the knowledge already found above, is as follows:

The area of the comb = the area of ivory + the area of holes; and $R_i = A_{i+1}$ for $i \geq 1$. Therefore (see Figure 3):

$$1 = (A_1 + 2A_2 + 4A_3 + 8A_4 + \cdots) + (1A_2 + 2A_3 + 4A_4 + \cdots)$$

from which

$$\begin{aligned}
1 - A_1 &= \tfrac{3}{2}(2A_2 + 4A_3 + 8A_4 + \cdots). \\
\text{And so } A &= \tfrac{2}{3}(1 - A_1) + A_1 = \tfrac{1}{3}(2 + A_1) \\
&= \tfrac{1}{3}(2 + \tfrac{1}{\alpha}) = \alpha^2/3.
\end{aligned}$$

This, then, is the sum of the infinite sequence just given above. The total area of the holes is:

$$\begin{aligned}
H &= 1 - A = 1 - \alpha^2/3 \\
&= \alpha^2/3. \quad (\text{Check: } 1 = A + H = \tfrac{1}{3}(\alpha^2 + \alpha^{-2}) = \tfrac{1}{3} \times 3).
\end{aligned}$$

3.34 Herta's star jewel

We now give a description of the star jewel, which was to ward off evil spirits from Herta.

Figure 4. The star jewel fractal, phases 2,3 and 5

A fractal can be constructed either outside the supporting polygon, or inside the polygon [6]. The resulting two fractals are quite different. Herta's goldpoint shield was constructed outside an equilateral triangle; the star jewel is the interior fractal.

In Figure 4 above, the fractal is shown being constructed on the inside of an equilateral triangle, using the same motif as for the shield (see Fig. 1). We hope the reader will see why we call this fractal a star jewel; it has three striking points, each connected to a central figure which looks like a sparkling diamond.

We shall not give any further analysis of the figure, but merely invite the reader to admire it.

3.4 The Goldpoint Motif Triangle, and Pentagon Fractals

In this final Section we first analyse the goldpoint motif triangle, showing various ways by which it can be partitioned.

Then we take a regular pentagon, and study some of its goldpoint properties. We show how a fractal of pentagon fractals can be constructed within it, and point out one or two of the properties of this object.

3.41 Properties of the goldpoint motif bounding triangle

In Figure 5(a) below, the goldpoint motif AGCHB is shown, together with its bounding triangle ABC. (It was also shown in Figure 1, above.) This triangle partitions into two (108°, 36°, 36°) triangles, viz. AGC and BHC, which we call S−triangles, and a (36°, 72°, 72°) triangle, GHC, which we call a T−triangle. We shall use the convention S_i to describe an S-triangle drawn on a base line segment of length $1/\alpha^i$, $i = 0, 1, 2, \ldots$; and similarly use T_i for T-triangles drawn on such base line segments.

When making the analyses and calculations, we shall have recourse to the formulae given at the beginning of Section 2, and also to the following trigonometric relations.

θ	36°	72°
$\sin\theta$	$\sqrt{\alpha + 2}/(2\alpha)$	$(1/2)\sqrt{\alpha + 2}$
$\cos\theta$	$\alpha/2$	$1/(2\alpha)$
$\tan\theta$	$\sqrt{\alpha + 2}/\alpha^2$	$\alpha\sqrt{\alpha + 2}$

3.42 The goldpoint motif triangle, and some partitions of it

Figure 5(a) is used to demonstrate several partition properties of the goldpoint motif triangle. Figure 5(b) shows how the triangle can be partitioned by pentagrams and S-triangles of diminishing sizes and with sides $1/\alpha^i$. Various calculations and comments on these figures are given below the diagrams.

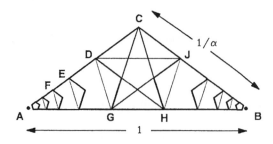

Figure 5(a). The motif triangle and some dividing lines

If $AB = 1$, then it is immediately seen that ABC is an S_0 triangle, which is partitioned by GC and HC into two S_{-1} and one T_{-3} triangles (since $AC = BC = 1/\alpha$ and $GH = 1/\alpha^3$). Thus $S_0 = 2S_{-1} \cup T_{-3}$.

The area of triangle ABC is $(1/2)AC\sin 36° = \sqrt{\alpha + 2}/(4\alpha^2)$.

Other partitions of ABC can be seen in the constructions. For example, the two $T-3$ triangles ADG and BJH together with the central pentagon $P-3$ on GH. Another is the set of decreasing and overlapping pentagons, on sides CD, DE, EF, ... and similarly on the right side of centre, whose union limitingly fills triangle ABC.

Finally, we observe that since an S-triangle can be partitioned into a T-triangle and an S-triangle (e.g. $ABC = AGC \cup GCB$), by repeated divisions ABC can be partitioned into a sequence of diminishing S-triangles; or else, similarly, into a sequence of diminishing T-triangles. We won't spell out their relative sizes, but point out that they are all in ratios of powers of α.

Figure 5(b). Pentagrams and S-triangles constructed in the motif triangle

Figure 5(b) demonstrates how the golden motif triangle can be partitioned into an attractive double sequence of diminishing pentagrams, with sides in diminishing powers of α, together with sequences of diminishing S-triangles.

3.43 Some properties of the regular pentagon, with goldpoints and partitions

The next two figures, 5(c) and 5(d), show regular pentagons, of side 1, with various construction lines upon them.

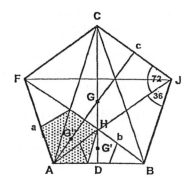
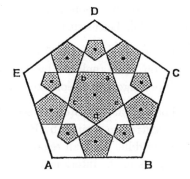

Figure 5(c). A regular pentagon   *Figure 5(d). A fractal of pentagons*

In fig. 5(c), $\triangle ACB$ is a T_1 triangle, so $AC = 1/(2\cos 72) = \alpha$. From $\triangle AGD$ we get $AG = 1/(2\cos 54) = \alpha/\sqrt{\alpha+2}$, and $GD = (1/2)\tan 54 = \alpha^2/2\sqrt{\alpha+2}$. And $CD = (1/2)\tan 72 = (1/2)\alpha\sqrt{\alpha+2}$.

By similar pentagons, $G'D = \alpha^{-3}GD$. And $G''A = \alpha G'D/\sin 54 = 1/(\alpha\sqrt{\alpha+2})$.

Proposition:

(i) $GG' = GG''$.

(ii) G is a goldpoint of CG'.

Proof:

$$
\begin{aligned}
(i) \quad GG' &= GD - G'D \\
&= \alpha^2/2\sqrt{\alpha+2} - 1/(2\alpha\sqrt{\alpha+2}) \\
&= 1/\sqrt{\alpha+2}, \text{ (since } \alpha^2 - 1/\alpha = 2).
\end{aligned}
$$

$$\text{and } GG'' = GA - G''A$$
$$= \alpha/\sqrt{\alpha+2} - 1/(\alpha\sqrt{\alpha+2})$$
$$= 1/\sqrt{\alpha+2}, \text{ (since } \alpha - 1/\alpha = 1);$$
$$\text{so } GG' = GG''.$$

$$(ii) \quad GG'/GC = GG'/GA$$
$$= 1/\sqrt{\alpha+2} \times (\sqrt{\alpha+2})/\alpha = 1/\alpha.\square$$

Other results about goldpoints in a pentagon construction may be found in [3, p. 28]. Let us turn to fig. 5(d), and examine the fractal of pentagons.

It is evident how fig. 5(d) can be obtained from fig. 5(c). The shaded pentagon P_2 is replaced by its inner pentagon (a P_4), and then the two small pentagons are replicated around pentagon abcde (a P_2).

Looking at fig. 5(d) we see $5P_3$s and $5P_4$s arranged alternately with their centres on a circle by Proposition (i) above, and with a pentagon P_2 in the middle. We can regard this as a motif for constructing a fractal of pentagons in the interior of pentagon $ABCDE$.

Thus, to arrive at phase 1, we must remove all points in the unshaded regions, together together with the perimeter of $ABCDE$. Then, to arrive at phase 2, we repeat the above constructions and removals in each of the eleven shaded pentagons. What remains will be 121 shaded pentagons, each scaled by a factor of α^i, $i = 2, 3,$ or 4 according to its construction. From the tree diagram below we see that the distribution of pentagons will then be: $1P_4, 10P_5, 35P_6, 50P_7, 25P_8$.

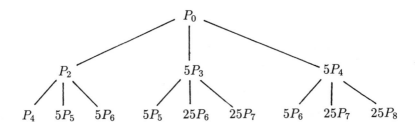

Evidently this process can be continued indefinitely. And formulae can be computed for the coefficients on the tree, and for reduction factors in areas when passing from phase i to phase $i + 1$.

The dust-set of the fractal is the set of points in $ABCDE$ which are not removed by this infinite process. A moment's thought shows that this set consists of the centres of all the pentagons constructed in the 'whole' process. And the set consists of a *cosmos* of points arranged in circles, with similar, reduced, circles arranged around each of them, and so on ad infinitum. Because of the similarity of this system with Ptolemy's model of the Universe, we name this dust-set the **Ptolemaic dust-set**.

3.44 Interior/Exterior goldpoint fractals

The next two figures show phases of the interior and exterior fractals which are constructed on a regular pentagon using the goldpoint motif on its sides.

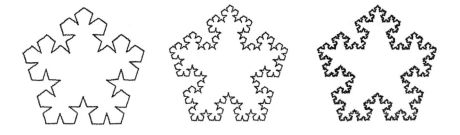

Figure 5(e). Interior goldpoint fractal of a pentagon, phases 2,3,4

Phase 2 shows an attractive clover-leaf arrangement of five leaves, each a portion of a P_1, around a central P_2.

Phase 4 shows clearly how the interior goldpoint fractal of a regular pentagon is equal to the exterior goldpoint fractal of its pentagram.

It is clear from the phase 1 diagram of figure 5(f) below, that the exterior goldpoint fractal of a regular pentagon is bounded by a regular pentagon. It is easy to prove this, using angle values of the S- and T-triangles which touch the boundary. We believe this property of a von Koch-type fractal having a bounding polygon which is similar to the generating polygon to be unique.

A final interesting comment is the following: the sharp points on the boundary in the phase 1 diagram can be connected by two unicursal polygons of chords of the diagram (each chord begins and ends along an arm of a point-angle); whereas the sharp points of the phase 2 diagram require four such unicursal polygons to join them all up. In phase n, there will be 2^n unicursal polygons required. The perimeters of the unicursals can be computed in terms of α, given that P_0 has side length 1. For example, in P_1, the two unicursals have perimeters 5 and $5(7 - 3\alpha)$ respectively.

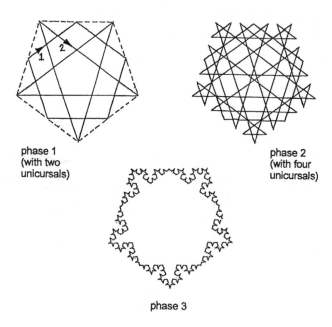

phase 1
(with two
unicursals)

phase 2
(with four
unicursals)

phase 3

Figure 5(f). Exterior goldpoint fractal of a pentagon, phases 1,2,3

Chapter 4

Triangles and Squares marked with Goldpoints

4.1 Introduction

It was explained in Chapter 1 how triangles marked with goldpoints first came to be studied, by Turner and the Atanassov family. We now take up that story again, and show how ideas about triangles and squares with goldpoints were developed. In particular, we will show how these objects (which we call *golden tiles*) can be fitted together, jig-saw fashion, in order to tile figures in the plane or in space.

Before studying tiling problems we will recall our definition of goldpoints, and prove two theorems about them in relation to two triangles forming a rhombus.

4.2 Equilateral Triangles and Goldpoints

In Part B, Section 1, Fibonacci vector polygons were studied, and it was shown how they behaved in the honeycomb plane. Each polygon zig-zags upwards, its sides lengthening and its points oscillating about, and tending towards, a limit line. They all tend to the same limit line.

It was shown that for each vector polygon, its vertices tend, as $n \to \infty$, to lie on the line $x/1 = y/\alpha = z/\alpha^2$, where α denotes the golden mean.

It is intriguing to observe how this line, as it leaves the origin and moves upwards, cuts sides of the vector polygons in ratios which involve α and the Fibonacci numbers.

The simplest cases are when the sides $[(1, 0, 1), (0, 1, 1)]$ and $[(0, 1, 1), (1, 1, 2)]$ are cut. These intersections occur as the limit line passes upwards from the equilateral triangle closest to the origin and through the triangle above it. The ratios in the two side-cuts are respectively $1 : \alpha$ and $\alpha : 1$. These observations constitute the *raison d' être* of Theorems 1 and 2, and of the triangular tiles to be described below. First we recall the definition of a goldpoint.

Definition: A *goldpoint* is an internal point of a line-segment, which divides the segment length in the ratio $\alpha : 1$ or $1 : \alpha$, where α is the golden mean $(1 + \sqrt{5})/2$.

Theorem 1. Let ABDC be a rhombus, with angle θ at A and D. Let AR be a ray through A which cuts CB in P and CD in Q.

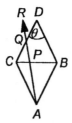

Then P is a goldpoint of CB if and only if Q is the goldpoint of CD such that $CQ/QD = \alpha$.
If $\theta = 60°$, then PC = QD.

Proof: Since $AB \parallel CD$, $\angle CQP = \angle PAB$ and $\angle QCP = \angle PBA$; and $\angle QPC = \angle APB$ (vertically opposite angles). Hence $\triangle PQC$ is similar to $\triangle PAB$.

Therefore $PB/PC = AB/QC = CD/CQ$ (since $AB = CD$).

If P is a goldpoint of CB (and by construction $CP < PB$), then $PB/PC = \alpha$. Hence $CD/CQ = \alpha$, so Q is a goldpoint of CD: note that $CD/CQ = (CQ + QD)/CQ = 1 + QD/CQ$. Therefore $QD/CQ = \alpha - 1 = 1/\alpha$, hence $CQ/QD = \alpha$.

For the converse, suppose that $CQ/QD = \alpha$.
Then $PB/BC = CD/CQ = 1 + 1/\alpha$, hence P is a goldpoint of CB.

Finally, if $\theta = 60°$ then $\triangle BCD$ is equilateral, so $CB = CD$. Then since $CB/PB = CD/CQ = \alpha$, we have $PB = CQ$ and hence $PC = QD$.   □

The following theorem shows how the 60°-rhombus can be set in the honeycomb plane, and 3D coordinate geometry used to determine the line through AR.

> **Theorem 2:** Let the rhombus of Theorem 1, with $\theta = 60°$, be set in the plane $x+y-z = 0$, with points A, B, C, D being respectively
> $(0,0,0), (1,0,1), (0,1,1), (1,1,2)$. Then if P is a goldpoint of CB, the ray AR lies in the line $x/1 = y/\alpha = z/\alpha^2$.

> **Proof:** In the accompanying diagram, $u \in (0, \frac{1}{2})$ is the fractional distance of P from C to B.

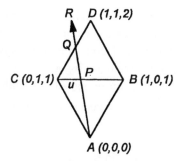

Figure 1. A 60°-rhombus in the plane $x + y - z = 0$

Since P is a goldpoint, and AR is to the left of AD, we have $PB/PC = \alpha$. Therefore:

$$PB^2 = \alpha^2 PC^2$$
$$(1-u)^2 + (u-1)^2 = \alpha^2(u^2 + u^2 + 0)$$
$$\Rightarrow \quad u - 1 = \pm u\alpha \quad \text{(reject +, since } u > 0\text{)}$$
$$\text{Hence} \quad u = 1/\alpha^2.$$

The coordinates of P are $(1/\alpha^2, 1-1/\alpha^2, 1) = 1/\alpha^2(1, \alpha, \alpha^2)$.

Therefore the line through AR has direction ratios $1 : \alpha : \alpha^2$; and the equation of the line follows immediately, since A is $(0,0,0)$. The direction cosines of the line are $(1, \alpha, \alpha^2)/(2\alpha)$, since $1 + \alpha^2 + \alpha^4 = 4\alpha^2$

$\square$

4.3   Golden Tiles and Combination Rules

The above theorems show that adjacent equilateral triangles, with gold-points marked on their sides, can give rise to interesting geometry involving the golden mean α. Having discovered this, Turner made up (in cardboard) a set of equilateral triangles, and he marked upon each side of every triangle one of its goldpoints. He proposed the following rules of equivalence and combination for the triangles (and, by extension, for polygons):

Rule 1. Two marked triangles in a plane are *equivalent* if one triangle can be rotated, and translated, until it and its three side-markings* coincide with the other triangle and its side-markings.

Rule 2. Two marked triangles can be *combined* (juxtaposed) if they can be placed in a plane with two sides coinciding and with the goldpoints in those two sides coinciding too. (Note: This is the *jigsaw combination move*; we shall speak of *jigging* two triangles together.) If we wish to indicate that two triangles, say T_1 and T_2, can be jigged, we shall write $T_1 * T_2$.

These same rules are to apply generally, to polygons.

4.31 Definitions of names

Golden tiles: Polygons with sides each marked with a goldpoint will be called *golden tiles*. The abbreviations TGP and SGP will be used for golden tiles made respectively from equilateral triangles and squares. These are abbreviations for 'triangles with goldpoints' and 'squares with goldpoints' respectively.

Tile figure: The *tile figure* of a golden tile with one mark per side is the polygon which is obtained by drawing on the tile straight lines which join up all pairs of marks occuring on adjacent sides of the tile.

Jigsaw: When two or more golden tiles are combined by jigsaw combination moves, the result will be called a *jigsaw* and the pattern displayed upon it, resulting from all the tile figures, will be called the *jigsaw pattern*.

4.32 Problems

Many types of problem can be posed about golden tiles and jigsaws. Combinatorial and geometric ones include: How many different (i.e inequivalent)

---

*[In this chapter, only one goldpoint per side will be marked. In general, however, both goldpoints may be marked on any side of a triangle, or polygon. Rules 1 and 2 cover the general case.]

tiles can be derived from a given polygonal shape? How many different ways can 2 (or 3, or 4 etc.) tiles be combined to form a jigsaw in a plane (or on the surface of a given polyhedral solid)? What kinds of tile figures are there; and what are their geometric properties (for example, what are their symmetries)? What kinds of jigsaw patterns can be made with a given set of golden tiles? The list could go on and on.

In this book we have space only to show a few of the results we have obtained, and those only for TGPs and SGPs. We begin by showing the tile figures for TGPs, and discussing a few of their properties.

4.4 The TGPs, and some Basic Properties

If one thinks of marking one goldpoint on each side of an equilateral triangle, since there are two different goldpoints on each side, there are 2^3 ways of placing the marks. However, the eight resulting TGPs are not all inequivalent. In fact, there are four equivalence classes, each with two TGPs in them. The set of four inequivalent TGPs is shown in Figure 2.

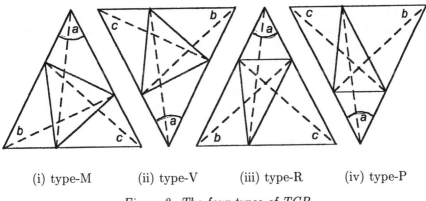

| (i) type-M | (ii) type-V | (iii) type-R | (iv) type-P |

Figure 2. The four types of TGP

It will be noted that the four tile figures in the TGPs are of two types, namely: an equilateral triangle, and a scalene triangle. The two tiles with equilateral triangles are inequivalent, and likewise are the two with scalene triangles. Evidently, tiles with equilateral and scalene triangles are inequivalent.

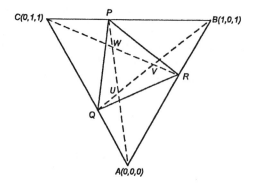

Figure 3. A TGP with full and dotted lines, labelled for Theorem 3

Rotations: When studying tile figures and jigsaw patterns, it will sometimes be necessary to specify rotations of TGPs in the plane. We define a
positive rotation to be an anti-clockwise motion, and a negative rotation to
be a clockwise one. Sometimes the vertices of a golden tile will be labelled,
so that rotations and combinations can be explicitly defined.

Turner coloured his original golden tiles Mauve (M), Violet (V), Red (R)
and Pink (P); hence the letters used in Fig. 2 to name these TGP types.
He also drew dotted lines on them, joining all vertices to the goldpoints on
sides opposite them. These added to the geometric complexity of the tile
figure, of course: Fig. 3 above shows one of these figures, and a theorem
about its properties follows.

Theorem 3. Let ABC be an equilateral triangle, with P,Q,R
goldpoints of the sides, marked such that $\triangle PQR$ is equilateral. Let AP meet BQ in U, and CR in W; and let CR meet
BQ in V. Then:

(i) U, V, W are mid-points of AP, BQ, CR respectively.

(ii) U, V, W are goldpoints of AW, BU, CV respectively.

(iii) Triangle areas: $\triangle UVW = \frac{1}{4}\triangle PQR = \frac{1}{2\alpha^4}\triangle ABC$.

Proof: The 3D coordinate system and lower triangle, explained for Theorem 2, is used again here. We study only ratios, or relative figures, so that the results are general (i.e. independent of the choice of side length $(\sqrt{2})$ for the triangle). We could use well-known vector geometry techniques to find the coordinates of **P**, **Q**, **R**, **U**, and **V**, and by these means effect proofs directly. However, we shall use a mixture of Euclidean and vector geometry, to demonstrate interesting ways for studying tile figures.

(i) Using 3D vector geometry, we find the points **P**, **U** and **W**, thus: $\mathbf{P} = (\alpha\mathbf{C} + \mathbf{B})/(\alpha + 1) = (1/\alpha^2, 1/\alpha, 1)$; similarly $\mathbf{Q} = 1/\alpha^2(0, 1, 1)$. Then $\mathbf{U} = \frac{1}{2}(1/\alpha^2, 1/\alpha, 1)$ is found from intersection of lines AP, QR, by standard procedures with coordinates. Hence **U** is the mid-point of AP, since A is $(0,0,0)$. Similarly, **V** and **W** are mid-points of BQ and CR.

(ii) From their coordinates, we see immediately that $\mathbf{W} = \alpha\mathbf{U}$. Then $AW : AU = \alpha : 1$, and hence U is a goldpoint of AW. Similarly V, W are goldpoints of BU, CV respectively.

(iii) The following argument leads to the ratio of areas of triangles PQR and ABC.

$$
\begin{aligned}
\Delta PQR &= \Delta ABC - 3\Delta AQR, \\
&\quad (\text{since } \Delta QPC \equiv \Delta PBR \equiv \Delta AQR)\,. \\
\Delta AQR &= (1/\alpha)\Delta AQB = (1/\alpha)((1/\alpha^2)\Delta ABC)\,. \\
\Rightarrow \quad \Delta PQR &= (1 - 3/\alpha^3)\Delta ABC = (2/\alpha^4)\Delta ABC\,.
\end{aligned}
$$

Next we find triangle UVW as a ratio of triangle ABC, thus:

$$
\begin{aligned}
\Delta UVW &= \Delta PQR - 3\Delta QVR\,. \\
\text{Now } \Delta QVR &= (1/2)\Delta QBR \quad (V \text{ is the mid-point of } QB) \\
&= (1/2\alpha^2)\Delta AQB = (1/2\alpha^2)(1/\alpha^2)\Delta ABC\,. \\
\Rightarrow \Delta UVW &= (1/2\alpha^4)\Delta ABC\,.
\end{aligned}
$$

Putting the results for ΔPQR and ΔUVW together, the proof of (iii) is completed. $\square$

It is of interest to note that we can cast part (i) of Theorem 3 in a way similar to that of Theorem 2, where a ray AR was involved. Thus:

Theorem $3'$(i). Let ABC be equilateral; and let AR_1, BR_2 be rays rotating about A and B respectively, with AD always kept equal to CE. Then when their intersection X is mid-point of AE, or of BD, E is a goldpoint of BC and D is a goldpoint of AC.

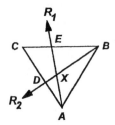

4.5 Forming Jigsaws, and Tiling with TGPs

We shall now discuss various results about how the TGPs can be combined using the jigsaw rule. We shall first treat combinations of two tiles: and then give examples of how TGPs can be used to tile the plane. Later several examples of jigsaw tiling of polyhedra surfaces by TGPs will be given. Because of lack of space, we can only give diagrams of these tilings, and not discuss the many aspects of their jigsaw patterns which beg to be studied and described.

We shall, however, begin with a few definitions of concepts which are involved in jigsaw formations, so that the reader will appreciate how rapidly the complexity of these combined objects develops. At every step in the development, new combinatorial problems of increasing difficulty suggest themselves.

4.51 Orientation and labelling of TGPs

Any jigsaw move in a plane can be carried out by a rotation, followed by a translation without rotation. In order to specify rotations of a TGP, it is necessary to have a fixed line in the plane (say H, which we shall say is 'horizontal') and refer the TGP to that line in a well-defined manner. Before saying how the fixed line becomes useful, we shall define an 'n-jigsaw' and a 'connected n-jigsaw', thus:

> **Definitions:** An *n-jigsaw* is a jigsaw composed of n tiles in a plane, combined by jigsaw moves. It is a *connected n-jigsaw* if every tile in it has at least one side (or point) coinciding with a side (or point) of another of the n tiles. Two n-jigsaws are *equivalent* if they can be translated and rotated in the plane until their jigsaw patterns are identical.

Labelling of TGPs: In Figure 2 we showed the four possible TGPs, and identified each by a letter (standing for a colour). We also labelled the three vertices of each tile, using the letters a, b and c, with anti-clockwise alphabetic ordering around the triangle in each case.

It may be noted that we always drew a triangle with one side horizontal (parallel to line H); theorems 4 and 5 below tell us that with any n-jigsaw of TGPs we can always rotate it in the plane until every one of its tiles has a horizontal side. Then, in each tile there will be one vertex either above its horizontal side or below it. Suppose in a particular tile T (where T is one of the four types M, V, R or P) this vertex has label l (where l is a, b, or c). Then we propose the following notation: T_l means that the l-vertex points downwards in T (it is below the horizontal side); whereas $T_{l'}$ means that the l-vertex points upwards in T (it is above the horizontal side). The two diagrams in Figure 4 illustrate this convention. They should convince the reader that the notation used, together with the labellings of TGPs given in Figure 2 above, specifies the orientation of tile T relative to line H precisely.

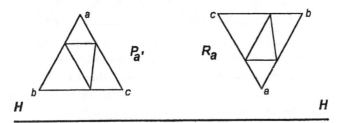

Figure 4. Two TGPs, labelled and oriented relative to line H

Theorem 4: If a tile T_1 has a horizontal side, and $T_1 * T_2$ is a 2-jigsaw, then tile T_2 has a horizontal side.

Proof: If T_2 is jigged to T_1 on the horizontal side of T_1, then T_2 has a horizontal side which coincides with that. If T_2 is jigged to one of T_1's sloping sides, then the 2-jigsaw is in the form of a rhombus: this is a parallelogram, one of whose sides is horizontal and *not* in T_1. Hence this horizontal side is in T_2. □

Theorem 5: Let a connected jigsaw (say J_n) be composed of n TGP tiles. Then if any one tile (say T_1) has a horizontal side, each of the tiles in the n-jigsaw has a horizontal side.

Proof: Proof is easily established by induction, using Theorem 4. □

4.52 Linear TGP jigsaws

A jigsaw composed of TGPs and which takes one of the following four forms will be called a linear TGP jigsaw, and designated an LTGP. Its *length* will be the number of TGPs used in its formation.

Linear forms: (i) $\triangle \triangledown \triangle \cdots \triangledown$ (ii) $\triangle \triangledown \triangle \cdots \triangle$
               (iii) $\triangledown \triangle \triangledown \cdots \triangledown$ (iv) $\triangledown \triangle \triangledown \cdots \triangle$.

Example

The following LTGP of length 4 is composed of one TGP of each type:

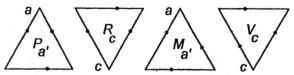

The symbol string $P_{a'} * R_c * M_{a'} * V_c$ gives a proper definition of one of these linear forms, and shows that one such actually exists. Moreover, since $V_c * P_{a'}$ is a valid combination, it follows that this LTGP can be extended indefinitely (in both directions). Hence we can say that this 4-jigsaw, of type defined by its symbol string, *tiles the jigsaw line*.

We can now ask the question: Does this LTGP tile the plane? The answer is: Yes. Two of these tile-strings may be placed end-to-end, or on top of or under one another, in suitably staggered positions. And the process may be continued indefinitely. These remarks could be made precise, using the symbols developed above; but we will leave it to the reader's visual imagination to confirm them.

4.53 Jigsaws with two tiles

Having studied single TGPs, and linear combinations of them, albeit most briefly, we can look at 2-jigsaws. Two jigged equilateral triangles form a 60° rhombus. We shall say that two of these rhombuses, each formed from two TGPs, are equivalent if their overall geometric patterns are identical (can be placed exactly over each other).

We ask how many inequivalent rhombus tiles there are, and the answer is 16. We have not space to give diagrams of all these; we present only two, in Fig. 5, to demonstrate how varieties of geometric patterns are presented on these rhombuses. Under the diagram we give a table which shows how often the various V, M, P and R tiles can be jigged together in pairs, in a total of 16 inequivalent ways.

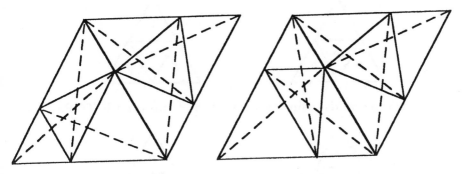

Figure 5. The two (M,P)-rhombuses

No. Pairs	V	M	P	R	Totals
V	0	1	1	2	4
M		0	2	1	3
P			2	5	7
R				2	2
Total					16

Table: Numbers of possible rhombuses, from TGP-pairs

As final comments on tiling the plane, we note that neither the V- nor the M-type tiles combine with themselves in pairs, so they cannot tile the plane. Although there are two types of (R,R) combinations, it can be shown that the R-tiles cannot tile the plane. Similarly, the pink tiles cannot tile the plane. Some of the rhombuses will tile the plane (e.g. $P_b * R_{a'}$ will do so); the (V,P) rhombus will not.

4.54 Tiling surfaces of solids

V. Assanova has discovered many examples of tilings of polyhedra. A small selection of her findings are presented below. Chapter 6 has more.

Each of the following diagrams is a developed, planar view of a surface tiling for the solid which is named above it. If these were to be cut out and folded, all the boundary edges could be matched in pairs, with goldpoints coinciding on these edges; and the solid's surface would result.

(A) Two pyramids, each using four TGPs of different types

 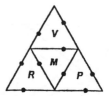

(B) Two pyramids, using two type-P and two type-R TGPs

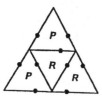 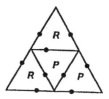

(C) A pyramid using four TGPs of each type

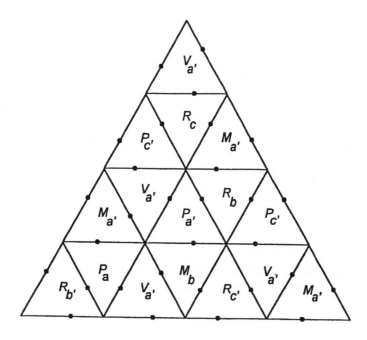

(D) An octahedron using two TGPs of each type

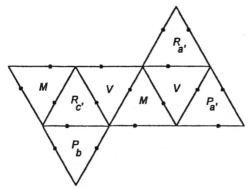

(E) A regular icosahedron using five TGPs of each type.
[N.B. Neighbouring faces are everywhere coloured differently.]

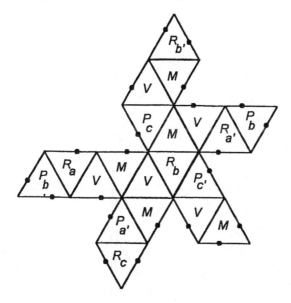

4.6   On Squares with Goldpoints (SGPs)

V. Atanassova extended the idea of goldpoint triangles to squares marked with goldpoints (SGPs), and she has discovered much about them. Not only has she studied some of their combinatorial and geometric problems, but she has also invented intriguing games which make use of them (see Chs. 6 and 7).

We shall first show the different types of SGP that can be drawn; and then demonstrate a few solutions to the problem of tiling the plane with them.

4.61 The six different (inequivalent) SGPs

The four goldpoints of a square may be marked in $2^4 = 16$ ways, since there are two ways to place a goldpoint on each of its sides. Allowing for plane rotations, there are six equivalence classes, which may be described in terms of the tile figures. These figures are: rectangle (1 type), square (2 types), trapezium (1 type) and quadrangle (2 types). A set of distinct representatives of these types is shown in Figure 6. (Note, incidentally, that they are arranged so that they jig-tile a 2×3-rectangle: and observe the centrally placed kite and diamond shapes, which play roles in SGP jigsaw patterns.)

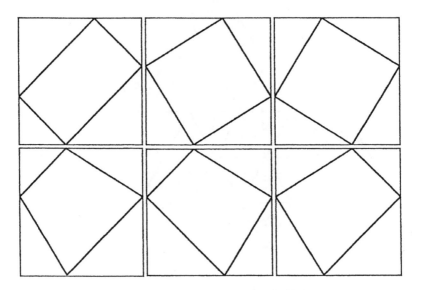

Figure 6. The six possible types of SGP

4.62 On tiling the plane

The four diagrams in Figure 7 below, each of four SGPs arranged in a 2×2 square, demonstrate how the plane may be tiled by SGPs in the following ways:

(i)   Using only the rectangle tile;
(ii) Using only the trapezium tile;
(iii) Using both types of quadrangle;
(iv) Using both types of square and two trapezia.

In each case, it is obvious how each 2×2 jig-tile can be jigged with a copy of itself, on either side, or on top or bottom; and thence how the plane can be tiled by further copies.

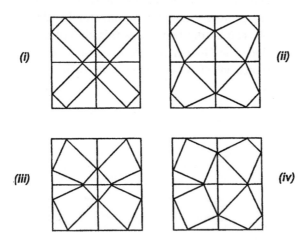

Figure 7. Four ways to tile the plane with SGPs

There are other solutions for jigging SGPs into 2×2 squares, some of which tile the plane, and some which do not. (N. B. Another plane 2×2 tiling solution may be found on the right side of Figure 6.)

4.7   Summary

The first part of the chapter was concerned with two theorems about rhombuses with a line being drawn from a vertex to cut two sides in goldpoints.

Then definitions of golden tiles, tile figures and jigsaws were given, followed by a theorem on properties of a type-V tile figure.

The rest of the chapter dealt with golden tiles which were either equilateral triangles or squares marked with one goldpoint on each of their sides. A variety of jigsaw problems were described, and solutions given.

The construction of golden tiles, and their applications to jigsaw tiling of plane and solid figures, provides a potentially endless source of fascinating problems in combinatorics and geometry. More are treated in the next two chapters. All results from these are of necessity, in view of the manner of tile markings, directly related to the golden mean.

In his book *The Divine Proportion*, Dover (1970), H. E. Huntley has given many geometric instances of what he calls the 'ubiquity of the golden mean'. Several relate to the angle 36° or multiples of that, for example when α arises in the geometry of a pentagram star (p. 28). Our relations of α to the angle 60° can hardly be new, as this kind of geometry has been studied at least since the Golden Age of classical Greece. However, we believe that our treatment of them, followed by their use in the study of golden tiles, is new.

Chapter 5

Plane Tessellations with Goldpoint Triangles

5.1 Introduction

In this chapter we shall begin by extending our jigsaw tile labelling techniques. Then we shall define jig-chains of TGPs, and show how the adjacency matrix for these chains can be used to study and count the ways of forming certain n-tiles. Much use will be made of Burnside's theorem for counting equivalence classes.

In particular the table given in Ch. 4 for counting 2-tiles will be verified, and then the 3-, 4-, 5- and 6-tiles will be counted. The general solution to the problem of tiling a hexagon with TGPs will be presented, and the hexagon tile-figures investigated.

5.2 The ls-String Notation for TGPs and SGPs

The diagrams in Fig. 1 show how we can describe the perimeter of a TGP-tile (or an SGP-tile) by recording the steps (long $= l$ or short $= s$) when passing (walking) around it, from vertex to goldpoint to vertex in the anticlockwise direction. We need only record the first step on each side, since the second must be its complement. This greatly simplifies all discussions and proofs about properties and relations between goldpoint jigsaw tiles.

First we must decide where we start the 'walk' around the tile, when measuring the steps. We require the start of each walk to be always from the left-hand vertex on the horizontal side of the TGP when it is 'horizontal side up' (HSU). When the TGP is 'horizontal side down' (HSD), we must start the walk at the right-hand vertex of the horizontal side. We shall use the same device with goldpoint squares (SGPs): the walk start on an SGP will always be from the left-hand vertex of the top edge of the square.

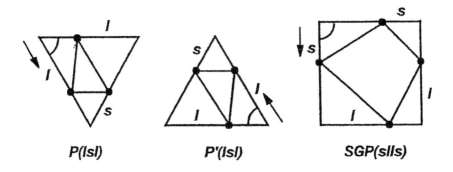

Figure 1. Examples of ls-labellings

We call the start vertices of walks *walk-origins*. In all future diagrams we shall mark the TGPs and SGPs (when necessary) with an angle sign, as shown in Figure 1.

> **Definition:** the succession of l, s labels of alternate steps around a golden tile, from its walk-origin, is called the *ls-string* of the tile.

It must be noted that the *ls*-string of a tile is measured on a particular aspect of the tile relative to a horizontal line in the plane. It is not a fixed attribute of the TGP concerned. We saw in Figure 1 that the *ls*-strings for P_a and $P_{a'}$ were the same; but for the other rotational aspects of the P-tile they are not.

The following diagram shows what happens to the *ls*-strings when the pink tile is rotated anti-clockwise through multiples of $60°$.

$$P_a \quad \rightarrow \quad P_{b'} \quad \rightarrow \quad P_c \quad \rightarrow \quad P_{a'} \quad \rightarrow \quad P_b \quad \rightarrow \quad P_{c'} \quad \rightarrow \quad P_a$$
$$(lsl) \qquad (sll) \qquad (lls) \qquad (lsl) \qquad (sll) \qquad (lls) \qquad (lsl)$$

We note that, given $P_a = lsl$, we can get $P_b (= P_{b'})$ and P_c by cyclically permuting lsl one element at a time, taking the first element to the third position and so on.

The ls-strings are very useful for proving various jigging properties of tiles. Several basic observations about them are the following: their proofs are immediate, from definitions.

> **Lemma:** If T_x is a TGP in HSU position (vertex x pointing down) and $T_{x'}$ is the TGP in HSD position (vertex x pointing up), then the ls-string is the same for both T_x and $T_{x'}$.
>
> **Theorem 1:** When two tile sides are placed next to each other, they will have opposite walk directions (see the diagram below Theorem 2).
>
> **Theorem 2:** A side of a unit TGP (or other golden tile) will jig to a side of another unit golden tile if and only if the ls-values of the two adjacent sides are complementary (see the diagram below). (N.B. The complement of l is s, and vice versa.)

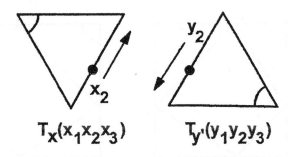

$$T_x(x_1x_2x_3) \qquad T_{y'}(y_1y_2y_3)$$

> **Theorem 3:** Let T_x be a TGP tile (HSU) with ls-string $x_1x_2x_3$; and T_y be a TGP tile with ls-string $y_1y_2y_3$. Then:
> (i) $T_y = T_{x'}$ (a triangle rotation of T_x through $180°$), if and only if $y_1y_2y_3 = x_1x_2x_3$ (N.B. $T_{x'}$ has the same ls-string as T_x.)   (ii) If T_y is a $60°$ anti-clockwise rotation of T_x, then $y_1y_2y_3 = x_2x_3x_1$ (N.B. This implies that $y_1 = x_2$, $y_2 = x_3$, $y_3 = x_1$.)
> (iii) If T_z is a $120°$ anti-clockwise rotation of T_x, then $z_1z_2z_3 = x_3x_1x_2$ Hence $z_1 = x_3$, $z_2 = x_1$, $z_3 = x_2$.

Finally we give a theorem about the jigging of two faces from squares with goldpoints (SGPs).

Theorem 4: Two faces (say G and H) from two unit cubes, which are formed from unit SGPs, will jig as $G * H$ if the complement of the ls-string of H is equal to $[(13)(2)(4) \times (ls\text{-}$ string of $G)]$, where (13) indicates a transposition of 1st and 3rd terms of G's ls-string.

Proof: The 1st side of G has to jig with the 3rd side of H, and vice-versa. Whereas the 2nd and 4th sides of G have to jig respectively with the 2nd and 4th sides of H (these are the horizontal sides, and have opposing directions when the pairs are jigged). All pairs of jigged sides have opposing directions, and the condition of Theorem 2 requires complementation.
□

Example: The diagram shows two SGPs, namely $G(lssl)$ and $H(llss)$. Applying the theorem, we find that they will jig if G is a face of one cube, and H a face of another, since $H(\overline{llss}) = (ssll) = (13)(2)(4) \times G(lssl)$.

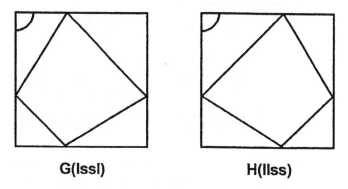

G(lssl)   **H(llss)**

5.3   Jigsaws with n = 2,3,4,5,6 TGPs

In Chapter 4 we dealt briefly with 2-tile rhombuses ($n = 2$ TGPs). Here we begin to study jigsaws more generally. The goal of this subsection is to reach a point where the tiling of a hexagon, using 6 TGPs, can be solved in some detail. We shall first discuss the positional aspects of TGPs in jigsaws, and then introduce a powerful counting tool, namely Burnside's Theorem, and use it to verify that there are only 4 types of TGP which are jigsaw inequivalent. Later we shall use it several times to count n-tile jigsaws.

5.31 The two positional aspects of TGPs in jigsaws

In view of Theorems 4 and 5 of Chapter 4, we know that all tiles in a jigsaw are either in the position ∇ (horizontal side up, HSU), or they are in position Δ (horizontal side down, HSD), given that one tile has a horizontal side. We shall confine ourselves to jigsaws of this kind, since all jigsaws of TGPs are rotationally equivalent to ones of this type.

Clearly the two triangles ∇ and Δ are not translation-equivalent, but they are rotational-equivalent. When constructing jigsaws with TGPs, both kinds of triangle must be taken into account. To be precise, we must consider the union of both of the following sets:

$$\text{HSU-triangles:} S = \{V, M, P_a, P_b, P_c, R_a, R_b, R_c\}$$
$$\text{HSD-triangles:} S' = \{V', M', P_{a'}, P_{b'}, P_{c'}, R_{a'}, R_{b'}, R_{c'}\}.$$

Let $S^+ = S \cup S'$. Then $|S^+| = 16$. Note that $V \neq V'$ and $M \neq M'$.

It is evident that a rotation of any of the triangles in S^+, through any multiple of $60°$, will carry it to itself or to another triangle in S^+. If a triangle is in S, a rotation through $60°$ will carry it to a triangle in S', and vice versa. It is easy to see that applying the group of rotations through multiples of $60°$ to all elements S^+ produces a group of permutations on the elements of S^+. The equivalence relation induced by this group has four equivalence classes, which are immediately found to be

$$\{V, V'\}, \ \{M, M'\}, \ \{P_a, P_b, P_c, P_{a'}, P_{b'}, P_{c'}\}, \ \{R_a, R_b, R_c, R_{a'}, R_{b'}, R_{c'}\}.$$

That is why we declared in Chapter 4 that there are four types of TGP, and designated them V, M, P, R.

It will be appropriate now to state Burnside's theorem*, which we shall use frequently later, and to illustrate its use by applying the permutation group to S^+ directly, thereby deducing that there are 4 distinct types of TGP.

---

*This theorem is a fundamental lemma in Pólya's Enumeration Theory, treated in most advanced textbooks on Combinatorics.

5.32 Burnside's Theorem

> **Theorem (Burnside's):** The number of equivalence sets into which a set S is divided by the equivalence relation induced by a permutation group G of S is given by $\frac{1}{|G|} \sum_{\pi \in G} \psi(\pi)$, where $\psi(\pi)$ is the number of elements of S that are invariant under the permutation π (i.e. its number of fixed points).

Application to S^+: Let the triangles in S^+ be ordered and labelled 1–16. If we apply the cyclic group of rotations through multiples of 60' to the triangles, we obtain a group of permutations, of order 6. Let us designate this by $G\{\pi_1, \pi_2, \pi_3, \pi_4, \pi_5, \pi_6\}$ To apply Burnside's theorem, we have to count the fixed points when applying each permutation in G, in turn, to the 16 elements of S^+. The following table lists and counts these fixed points.

Rotation		#	Invariant maps
60°	(π_1)	0	
120°	(π_2)	4	$V \to V, V' \to V', M \to M, M' \to M'$
180°	(π_3)	0	
240°	(π_4)	4	$V \to V, V' \to V', M \to M, M' \to M'$
300°	(π_5)	0	
360°	(π_6)	16	Identity permutation
Total:		24	

Hence, applying Burnside's theorem, the number of different TGP types (i.e. the number of equivalence classes) is $(1/6) \times 24 = 4$.

This theorem makes the counting of equivalence classes relatively simple, since only the invariant mappings, or fixed points of the permutations, have to be determined.

5.33 Notations for the tiles in S

When considering how to jig TGPs together, it is necessary to think of eight different tiles as in S, or sometimes sixteen as in S^+, because we need to take account of their positions relative to the horizontal line H (see Figures 2 and 4 of Ch. 4). The following table introduces notations for the eight HSU-TGPs of set S.

We have assigned the digits $1, 2, ..., 8$ to these different (HSU-positioned) tiles. The three rows give the alphabetic, numeric, and *ls*-string notations for each of the eight tiles.

Alphabetic :	V	M	P_a	P_b	P_c	R_a	R_b	R_c
Numeric :	1	2	3	4	5	6	7	8
ls-string :	lll	sss	lsl	sll	lls	lss	ssl	sls

Table 1. The alphabetic, numeric and ls-string designations of TGPs

5.34 TGP n-chains

In Figure 2 below, we give an example of four TGPs, all 'horizontal side up' arranged in a horizontal line; and we call such an arrangement a 4-chain of TGPs. If they actually jig together properly, as they do in the example, we may place an asterisk between adjacent triangles.

In general we shall talk of *n-chains of TGPs*, when n TGPs are arranged in this way.

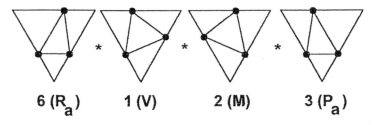

6 (R_a)   **1 (V)**   **2 (M)**   **3 (P_a)**

Figure 2. Example: a 4-chain of TGPs which will jig downwards

Using the numeric definitions of tiles, from Table 1, we can designate the whole chain merely by writing down the digit string 6123.

We developed this numeric notation in order to be able to write computer programs for solving combinatoric problems about TGPs, and also to have a shorthand method for recording jigsaws of several tiles. The reader will see much evidence of its use from now on.

5.35 The jigging matrix

We can check which TGP pairs can be jigged, in a 2-chain, and record all the successful pairs in a matrix, in the same way that a connection (or transition) matrix is computed for a directed graph. Thus, in the following matrix, a 0 in row i and column j means that the jig-pair $i*j$ is not possible; whereas a 1 in that position means that $i*j$ is possible. We shall designate the matrix M, and call it *the jigging matrix* for the set of oriented and labelled TGP-tiles.

Exactly as with directed graphs, we can multiply M by itself n times, and then the ij-th element of M^n will be an integer which counts the number of possible n-chains which begin with TGP labelled i, and end in TGP labelled j. (A proof of this can be found in any Graph Theory text-book.) Below we give the matrices M and M^2.

$$M = \begin{array}{c|cccccccc} & 1 & 2 & 3 & 4 & 5 & 6 & 7 & 8 \\ \hline 1 & 0 & 1 & 0 & 1 & 0 & 0 & 1 & 1 \\ 2 & 1 & 0 & 1 & 0 & 1 & 1 & 0 & 0 \\ 3 & 1 & 0 & 1 & 0 & 1 & 1 & 0 & 0 \\ 4 & 0 & 1 & 0 & 1 & 0 & 0 & 1 & 1 \\ 5 & 0 & 1 & 0 & 1 & 0 & 0 & 1 & 1 \\ 6 & 1 & 0 & 1 & 0 & 1 & 1 & 0 & 0 \\ 7 & 1 & 0 & 1 & 0 & 1 & 1 & 0 & 0 \\ 8 & 0 & 1 & 0 & 1 & 0 & 0 & 1 & 1 \end{array}$$

$$M^2 = \begin{array}{c|cccccccc} & 1 & 2 & 3 & 4 & 5 & 6 & 7 & 8 \\ \hline 1 & 2 & 2 & 2 & 2 & 2 & 2 & 2 & 2 \\ 2 & 2 & 2 & 2 & 2 & 2 & 2 & 2 & 2 \\ 3 & 2 & 2 & 2 & 2 & 2 & 2 & 2 & 2 \\ 4 & 2 & 2 & 2 & 2 & 2 & 2 & 2 & 2 \\ 5 & 2 & 2 & 2 & 2 & 2 & 2 & 2 & 2 \\ 6 & 2 & 2 & 2 & 2 & 2 & 2 & 2 & 2 \\ 7 & 2 & 2 & 2 & 2 & 2 & 2 & 2 & 2 \\ 8 & 2 & 2 & 2 & 2 & 2 & 2 & 2 & 2 \end{array} = 2U$$

Successive multiplication by M gives:

$$M^3 = 2MU = 8U \; ; \; M^4 = 8MU = 32U \; ; \; \cdots \; ; \; M^n = 2^{2n-3}U.$$

5.36 Formation of rhombuses from 2-chains

If a 2-chain of TGPs is jigged downwards, the result is a 60° rhombus. It is immaterial whether the left-hand triangle is dropped, or the right-hand one, for the two rhombuses are jigsaw equivalent: a 60° rotation takes one into the other. In chapter four (p. 202) we gave a table showing how many different rhombuses can be formed from the four different types of TGPs.

The information in that table (and the jigging matrix M given in 5.35) was found by trial and error, moving cardboard triangles around like jigsaw pieces. It took some time to get all the counts right. The final, correct, count of inequivalent rhombuses is sixteen.

In figure 3 below, we give the jigging matrix M again, with a lot more information added to it; and beneath it we give the table of rhombus counts, blocked to colour-pairs only. After the figure, we shall explain how the new matrix M was obtained, and show how it may be used to confirm the rhombus count-table. Later we shall use the much easier method of Burnside's to re-confirm the total.

$M^+ =$

	TGP		V	M	P_a	P_b	P_c	R_a	R_b	R_c
	no.		1	2	3	4	5	6	7	8
	ls-string		lll	sss	lsl	sll	lls	lss	ssl	sls
1	V	lll	0	1_1	0	1_2	0	0	1_3	1_4
2	M	sss	1_1	0	1_5	0	1_6	1_7	0	0
3	P_a	lsl	1_2	0	1_8	0	1_9	1_{10}	0	0
4	P_b	sll	0	1_6	0	1_9	0	0	1_{11}	1_{12}
5	P_c	lls	0	1_5	0	1_8	0	0	1_{13}	1_{14}
6	R_a	lss	1_3	0	1_{13}	0	1_{11}	1_{15}	0	0
7	R_b	ssl	1_4	0	1_{14}	0	1_{12}	1_{16}	0	0
8	R_c	sls	0	1_7	0	1_{10}	0	0	1_{15}	1_{16}

(a) The jigging matrix, with added information (M^+)

	V	M	P	R
V	0	1	1	2
M	1	0	2	1
P	1	2	4	5
R	2	1	5	4

(b) Table 2. The blocked rhombus counts

Figure 3. Counting rhombuses

In the matrix M^+ we have shown the ls-string for each triangle, as measured from its top left-hand corner. From these strings, we can immediately fill in the 8×8 table of 0s and 1s, because we know that a 2-chain will jig if the second ls-symbol on the left-hand triangle (second column) is equal to the complement of the first ls-symbol on the right-hand triangle (top row). Otherwise it won't.

We have appended a subscript to each 1 in the table, and it may be checked that the subscripts $1, 2, 3, \ldots, 16$ occur twice each within the table. If a pair of TGPs form a rhombus associated with symbols 1_i in M, and another pair form a rhombus again with symbols 1_i in M, then the two rhombuses are rotationally equivalent. That is, they are the same jigsaw piece.

An example is 1^5 which occurs for both the jigging 2-chains $M * P_a$ and $P_c * M$. When jigged (dropping the left tile) they form the two rhombuses $\Delta\nabla = MP_a$ and $\Delta\nabla = P_{a'}M$ respectively. Since a 180° rotation will carry one into the other, they represent the same jigsaw piece. This means, immediately, that the 32 pairs indicated in M^+ form exactly 16 inequivalent rhombuses.

We can prove this fact directly using Burnside's theorem, thus:

Form the 32 rhombuses, one for each 1 in the matrix, and apply the rotation group $G(0^0, 180°)$ to each. Note that a 60°-rhombus has 2-fold rotational symmetry about its centroid, and that no rhombus pattern can rotate into itself by a 180° rotation.

Then the permutations on the 32 rhombuses induced by the 0° and 180° rotations have respectively 32 and 0 fixed points. Therefore, the number of rotationally inequivalent rhombuses is $(1/2)(32 + 0) = 16$.

5.37 Formation of jigsaw n-tiles, for $n = 3, 4, 5$

Now that we have dealt with 2-tile rhombuses in some detail, we shall proceed to count the 3-tiles, 4-tiles and 5-tiles which are made from TGPs using the jigsaw rules. Figure 4 below shows the various possible shapes for these tiles.

Counting the 3-tiles

There are only two ways of placing three TGPs together to form a 3-tile, with at least one side being horizontal. These provide the following two shapes: $\Delta\nabla\Delta$ and $\nabla\Delta\nabla$. But since a rotation through 180° will carry one of these shapes into the other, we only need consider and count the left-hand shape of jigsaw 3-tiles. This is easily done, as follows:

Let us think of the three triangles as A, B, C, in the order they appear in the jigsaw shape. Then there are 8 possible TGPs for filling A, then 4 for filling B (see the rows in the jigging matrix M), and again four for filling C. Thus there are $8 \times 4 \times 4 = 128$ different 3-tiles. Note that a 3-tile has no rotational axis of symmetry, so we do not need to consider fixed points of permutation operations here.

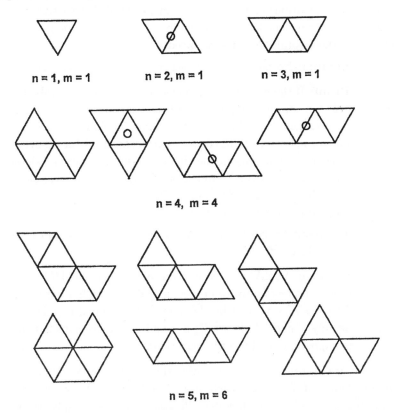

Figure 4. The possible shapes of 1-, 2-, 3-, 4- and 5-tiles of TGPs

Counting the 4-tiles

($n = 4$ row: labelled 4(a), 4(b), 4(c), 4(d) left to right in Fig. 4)

The linear forms:

We shall first count the two linear forms 4(c) and 4(d). We note first that these are not equivalent shapes under a 180° rotation: but it is evident that they will have the same numbers of inequivalent jigsaw pieces. So we need only count the type 4(c) pieces: there will be an equal number of type 4(d) cases.

Designate the four TGPs of $\nabla\triangle\nabla\triangle$ (in 4(c)) by A, B, C, D reading from the left. Then there are 8 choices of TGP for A, then four choices for each of B, C, D taken in turn: a total of 8×4^3 of 4-tiles of shape 4(c). However, under rotations of 180° in the plane, each of these 4-tiles rotates into another of them. The following theorem tells us that there are no fixed points under these rotations. First we need a lemma.

Lemma: No pair of TGPs such as $\nabla\Delta = X.X'$ can jig.

Proof: If the corners of the left-hand triangle X be labelled a, b, c in an anticlockwise direction, starting at the top-left corner, then X' will have anticlockwise labelling a, b, c starting at the bottom-right corner. This means that the adjacent sides of the triangles in the rhombus, taken in anticlockwise directions, are bc on X, and bc on X'. Hence both these sides have the same ls-value, and by Theorem 2 of this chapter, they do not jig on those sides. $\square$

Theorem 5: No linear 4-tile of TGPs can rotate through 180° into itself.

Proof: If $\Delta\nabla\Delta\nabla$ is to rotate into itself, then the TGP tiles used must be of only two types, jigged in the following form: $X' * Y * Y' * X$. But by the lemma, the central two TGPs, namely Y and Y', do not jig as required. Hence such a 4-tile is impossible, and the theorem is true. $\square$

Now we are in a position to determine the number of inequivalent linear jigsaw 4-tiles of TGPs (for each of 4(c) and 4(d)). We showed above that there are $8 \times 4^3 = 512$ possible tile combinations. Applying Burnside's theorem to these, with the group of rotations $G(0°, 180°)$ inducing an equivalence relation in them, we find that the number of equivalence classes amongst the 512 4-tiles is $(1/2)(512 + 0) = 256$. Here the 0 is the number of fixed points under 180° rotations, and is a consequence of Theorem 5.

The following theorem generalises this result.

Theorem 6: The number of distinct linear $(2n)$-tiles is 2^{4n}, of either kind.

Proof: By the same arguments as for the 4-tiles, there are $8 \times 4^{2n-1} = 2^{4n+1}$ possible linear $(2n)$-tiles, of either kind. There can be no fixed points under $180°$ rotations (applying the lemma to the centre two TGPs). Hence by Burnside's theorem, there are a total of $(1/2)(2^{4n+1} + 0) = 2^{4n}$ different linear $(2n)$-tiles, of either kind. $\qquad\qquad\square$

The shape 4(a):

Since there are no rotational symmetries for this shape, we find immediately that there are $8 \times 4^3 = 2^9$ ways of filling the triangles by TGPs, drawn from the 8 types available: we can fill the first one (any) in 8 different ways, and then each of the others can be filled in turn, laying each against an already filled one, in $4 \times 4 \times 4$ ways. Thus there are 512 different 4(a) tiles.

The shape 4(b):

Let us call this the *triangle 4-tile*. Again, as for the shape 4(a), there are 512 different ways to place TGPs within the four triangles. This time, however, some of these triangle 4-tiles have rotational symmetries, and these must be treated.

The triangle 4-tile shape has 3-fold rotational symmetry, with the group of rotations being $G(0°, 120°, 240°)$. For the second two rotations, only those whose inner TGP is either V or M, and with all three outer TGPs being the same as each other, can rotate into themselves — that is be rotation invariant (fixed points). So we can fill the inner triangle in 2 ways (V or M), and then choose any one of the four TGPs which will jig with V or M, as the case may be.

Hence there are $2 \times 4 = 8$ triangle 4-tiles which have 3-fold rotational symmetry. So the numbers of fixed points in the induced permutations of the 512 tiles is $512, 8, 8$ respectively for the three corresponding rotations $0°, 120°, 240°$ of the symmetry group G.

Finally, then, using Burnside's theorem we obtain the number of distinct jigsaw triangle 4-tiles as: $(1/3)(512 + 8 + 8) = 176$.

Counting the 5-tiles

The six possible shapes of the 5-tiles composed of TGPs are shown in
Figure 4 above. Since each shape has no rotational symmetries (except
1-fold), there can be no rotational fixed points when five TGPs are placed
into those shapes, except under the identity rotation. Hence the number of
possible 5-tiles of TGPs, of each of the four shapes, is: $8 \times 4^4 = 2^{11}$.

5.38 Studies of the 6-tiles of TGPs

Figure 5 below shows the nineteen different shapes which can be formed
from six jigged TGPs. These shapes are non-equivalent under translations
and rotations in the plane.

Figure 5. The possible shapes of 6-tiles

Note that most of these 6-tiles do not have rotational symmetries, other
than the trivial 1-fold one. Those that have are marked with small circles,
placed at the centres of rotational symmetry. Five of these have 2-fold rotational symmetry: but, in view of the lemma before Theorem 5, none of

these shapes can be rotated into themselves when filled with TGPs. Hence we do not have to consider their individual rotation properties when applying Burnside's theorem, except that we must use the group $G(0°, 180°)$ rather than the trivial one used for the others.

Counting the 6-tiles

To count the numbers of possible 6-tile jigsaws, filling any but the hexagon shape with selections of the 8 TGPs, we have only to apply the simple arguments given several times above for smaller n-tiles. Consider any 6-tile shape but the hexagon one. Choose an empty triangle, and fill it with any TGP chosen from the 8 available. Then move to an adjacent empty triangle, and (see the jigging matrix M) there are 4 TGPs available to fill it. Repeat this latter process until all six tiles are filled with a TGP. Then that procedure provides a total of $8 \times 4^5 = 2^{13} = 8192$ jigsaw tiles of the given shape.

Now, using Burnside's theorem on the 6-tiles with 1-fold symmetry, we obtain the result that there are $(1/1) \times 8192 = 8192$ distinct such jigsaw 6-tiles of each shape. Whereas for each of the shapes which have 2-fold symmetry, there are $(1/2)(8192 + 0) = 4096$ distinct jigsaw 6-tiles.

The one remaining shape, the 6-tile hexagon, has 6-fold rotational symmetry, and hence its possible jigsaw tiles are much harder to count. When it is filled with TGPs, fixed points under rotations can arise. We shall treat this complicated case in a separate subsection, to follow.

On the hexagons formed from six TGPs

We now show how to count the hexagon 6-tiles; then we determine how many of these are distinct jigsaw hexagons: the answer is shown to be 700. A study of the inner hexagons (tile figures in the 6-tile hexagons) is then described.

If we take a 6-chain of TGPs which will jig, and fold it downwards and around until all adjacent sides meet, we will get a tiled hexagon — provided that the first side of the first TGP will jig with the second side of the last TGP of the chain. An example of such a tiling follows:

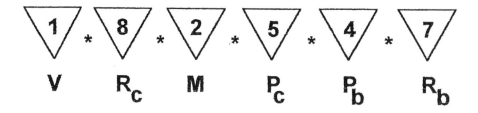

Figure 6. A 6-chain which jigs to a hexagon

We can easily count how many hexagons can be formed in this way, if we do not allow for repetitions due to rotational symmetries. We can do this by computing the 6th power of the jigging matrix M, and then adding the numbers on its leading diagonal.

Diagonal elements of the matrix M^6 count the 7-chains which begin with one TGP and end with the same TGP. That is, they count the 6-chains which will fold around and jig the sixth TGP with the first TGP (since the first TGP is congruent to the last TGP of the 7-chain). Now $M^n = 2^{2n-3}U$, hence $M^6 = 2^9 U$. There are 8 diagonal elements, so the total of these is $8 \times 2^9 = 2^{12} = 4096$.

[An easier way to count the hexagons is to multiply the ways of filling one triangle (from the 8 TGPs available), by the number (4) ways of filling an adjacent triangle, and so on, until the last triangle is reached: it can be filled in only 2 ways. The total is therefore $8 \times 4^4 \times 2 = 2^{12}$.]

Each of the above 7-chains has to be listed, and the first six TGPs (the 6-chain) of each has to be examined for cyclic permutational symmetries. For example, one hexagon-tiling is given by the chain 121212 (i.e. alternating Violet and Mauve TGPs). In the listing of 4096 possible tilings, the chain 212121 will also occur: this must be discarded from the count, since it codes the same hexagon tile as does 121212. A rotation through 60° will carry one into the other.

The task of sorting out the set of hexagon-tiling 6-chains into types, and then discovering all the redundant chains, was too large to tackle 'by hand' (although it was attempted!); so a computer program was written to do it. The set of inequivalent 6-chains which was obtained thus is listed in the table below. The count of this list, that is the number of hexagon

6-tiles, is 700. Below the list, we shall provide a mathematical proof for the number 700.

Table 3: The set of 6-chains which will tile a hexagon
$(1 = V, 2 = M, 3 = P_a, 4 = P_b, 5 = P_c, 6 = R_a, 7 = R_b, 8 = R_c)$

121212	121217	121233	121236	121252	121257	121263	121266
121423	121426	121442	121447	121473	121476	121482	121487
121717	121733	121736	121752	121757	121763	121766	121823
121826	121842	121847	121873	121876	121882	121887	123123
123126	123142	123147	123173	123176	123182	123187	123317
123333	123336	123352	123357	123363	123366	123523	123526
123542	123547	123573	123576	123582	123587	123617	123633
123636	123652	123657	123663	123666	125217	125233	125236
125252	125257	125263	125266	125423	125426	125442	125447
125473	125476	125482	125487	125717	125722	125736	125752
125757	125763	125766	125823	125826	125842	125847	125873
125876	125882	125887	126126	126142	126147	126173	126176
126182	126187	126317	126333	126336	127352	126357	126363
126366	126523	126526	126542	126547	126573	126576	127582
126587	126617	126633	126636	126652	126657	126663	126666
142142	142147	142173	142176	142182	142187	142317	142333
142336	142352	142357	142363	142366	142523	142526	142542
142547	142573	142576	142582	142587	142617	142633	142636
142652	142657	142663	142666	144217	144233	144236	144252
144257	144263	144266	144423	144426	144442	144447	144473
144476	144482	144487	144717	144733	144736	144752	134757
144763	144766	144823	144826	144842	144847	144873	144876
144882	144887	147147	147173	147176	147182	147187	147317
147333	147336	147352	147357	147363	147366	147523	147526
147542	147547	147573	147576	147582	147587	147617	147633
147636	147652	147657	147663	147666	148217	148233	148236
148252	148257	148263	148266	148423	148426	148442	148447
148473	148476	148482	148487	148717	148733	148736	148752
148757	148763	148766	148823	148826	148842	148847	148873
148876	148882	148887	171717	171733	171786	171752	171757
171763	171766	171823	171826	171842	171847	171873	171876
171882	171887	173173	173176	173182	173187	173333	173336

173352	173357	173363	173366	173523	173526	173542	173547
173573	173576	173582	173587	173633	173636	173652	173657
173663	173666	175233	175236	175252	175257	175263	175266
175423	175426	175442	175447	175473	175476	175482	175487
175733	175736	175752	175757	175763	175766	175823	175826
175852	175847	175873	175876	175882	175887	176176	176182
176187	176333	176336	176352	176357	176363	176366	176523
176526	176542	176547	176573	176576	176582	176587	176633
176636	176652	176657	176663	176666	182182	182187	182333
182336	182352	182357	182363	182366	182523	182526	182542
182547	182573	182576	182582	182587	182633	182636	182652
182657	182663	182666	184233	184236	184252	184257	184263
184266	184423	184426	184442	184447	184473	184476	184482
184487	184733	184736	184752	184757	184763	184766	184823
184826	184842	184847	184873	184876	184882	184887	187187
187333	187336	187352	187357	187363	187366	187523	187526
187542	187547	187573	187576	187582	187587	187633	187636
187652	187657	187663	187666	188233	188236	188252	188257
188263	188266	188423	188426	188442	188447	188473	188476
188482	188487	188733	188736	188752	188757	188763	188766
188823	188826	188842	188847	188873	188876	188882	188887
233335	233354	233358	233365	233525	233544	233548	233575
233585	233588	233635	233654	233658	233665	235235	235254
235258	235265	235425	235444	235448	235475	235484	235488
235735	235754	235758	235765	235825	235844	235848	235875
235884	235888	236335	236354	236358	236365	236525	236544
236548	236575	236584	236588	236635	236654	236658	236665
252525	252544	252548	252575	252584	252588	252635	252654
252658	252665	254254	254258	254265	254444	254448	254475
254484	254488	254735	254754	254758	254765	254844	254848
254875	254884	254888	257335	257354	257358	257365	257544
257548	257575	257584	357588	257635	257654	257658	257665
258258	258265	258444	258448	258475	258484	258488	258735
258754	258758	258765	258844	258848	258875	258884	258888
263335	263354	263358	263365	263544	263548	263575	263584
263588	263635	263654	263658	263665	265265	265444	265448
265475	265484	265488	265735	265754	265758	265765	265844
265848	265875	265884	265888	266335	266354	266358	266365
266544	266548	266575	266584	266588	266635	266654	266658

266665	333333	333336	333357	333366	333547	333576	333587
333636	333657	333666	335447	335476	335487	335736	335757
335766	335847	335876	335887	336336	336357	336366	336547
336576	336587	336636	336657	336666	354447	354476	354487
354736	354757	354766	354847	354876	354887	357357	357366
357547	357576	357587	357636	357657	357666	358447	358476
358487	358736	358757	358766	358847	358876	358887	363636
363657	363666	365447	365476	365487	365757	365767	365847
365878	365887	366366	366547	366576	366587	366657	366666
444444	444448	444475	444488	444758	444765	444848	444875
444888	447548	447575	447588	447658	447665	448448	448475
448488	448758	448765	448848	448875	448888	475475	475488
475758	475765	475848	475875	475888	476548	476575	476588
476658	476665	484848	484875	484888	487575	487588	487658
487665	488488	488758	488765	488875	488888	575757	575766
575876	575887	576576	576587	576666	587587	587666	588766
588876	588887	666666	888888				

Proof that there are 700 hexagon jigsaw tiles

We have already shown that there are $2^{12} = 4096$ different hexagon 6-tiles, when rotations are not considered.

If we wish to count the jigsaw-different hexagons, we must allow for rotations in the group $G(0°, 60°, 120°, 180°, 240°, 300°)$. In order to apply Burnside's theorem, we must imagine applying the six rotations to the set S of 4096 hexagon 6-tiles, and counting the invariant transformations (fixed points) among them.

Fortunately we need not check all combinations of six tiles. The task is considerably cut down by using four theorems (given below), selection of colour combinations, and the fact that only conditions on one, two, or three of adjacent tiles in a hexagon have to be checked if symmetry occurs.

The following Figure shows the five types of hexagon which have rotational symmetries. The last two are of different colour patterns, but they have the same 2-fold symmetry. The letters A,B,C refer to TGPs of three particular colours: note how they are arranged all facing the centre point of the hexagon. This means that any of these hexagons would be formed from the six types shown, by arranging them in a 6-chain, and then jigging them downwards. Which means that one can then refer to the jigging matrix M^+ to obtain their ls-strings directly.

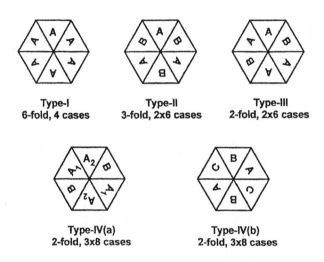

Figure 7. Hexagon tiles with rotational symmetries

The hexagons which have the symmetry types shown in the Figure can be fairly quickly found by means of the following theorems, and using the jigging matrix. Tree diagrams, drawn from each TGP and using the 1s in M, extending each tree down three levels, were found helpful to check that all the symmetric hexagon types had been found. Note that it is only necessary to check at most three conditions; and select appropriate colour combinations of one, two, three or four TGPs.

Theorem 7: A hexagon of type-I can be formed from six tiles $A(a_1, a_2, a_3)$ if and only if $a_1 = \bar{a}_2$.

Theorem 8: A hexagon of type-II can be formed from three tiles $A(a_1, a_2, a_3)$ and three tiles $B(b_1, b_2, b_3)$ if and only if
$$a_1 = \bar{b}_2 \text{ and } a_2 = \bar{b}_1.$$

Theorem 9: A hexagon of type-III can be formed from four tiles $A(a_1, a_2, a_3)$ and two tiles $B(b_1, b_2, b_3)$ if and only if
$$a_1 = \bar{b}_2, \ a_2 = \bar{b}_1 = \bar{a}_1.$$

Theorem 10: Types IV(a) and IV(b) are in the same category for counting fixed points, even though the former has two colours whilst the latter has three. Let us designate $A_1 = A$ and $A_2 = B$. Then: two tiles $A(a_1, a_2, a_3)$, two tiles $B(b_1, b_2, b_3)$ and two tiles $C(c_1, c_2, c_3)$ will form a hexagon of type IV if and only if $a_1 = \bar{c}_2$, $b_1 = \bar{a}_2$, and $c_1 = \bar{b}_2$.

The list of hexagons with rotational symmetries, together with their contributions to the fixed point counts (omitting the ones from the identity permutation) needed for Burnside's theorem is as follows (see Figure 7):

Type I: 4 cases, 5 fixed points each.

$$333333 \quad 444444 \quad 666666 \quad 888888$$

Type II: 12 cases, 2 fixed points each.

$$
\begin{array}{cccc}
121212 & 212121 & 171717 & 717171 \\
252525 & 525252 & 363636 & 636363 \\
484848 & 848484 & 575757 & 757575
\end{array}
$$

Type III: 12 cases, 1 fixed point each.

$$
\begin{array}{ccc}
336336 & 363363 & 633633 \\
366366 & 663663 & 636636 \\
448448 & 484484 & 844844 \\
488488 & 884884 & 848848
\end{array}
$$

Type IV: 48 cases, 1 fixed point each.

$$
\begin{array}{llll}
(b) & 123123 & 231231 & 312312 \\
(b) & 126126 & 261261 & 612612 \\
(b) & 142142 & 421421 & 214214 \\
(b) & 147147 & 471471 & 714714 \\
(b) & 173173 & 731731 & 317317 \\
(a) & 176176 & 761761 & 617617 \\
(b) & 182182 & 821821 & 218218 \\
(a) & 187187 & 871871 & 718718
\end{array}
$$

(a)	235235	352352	523523
(a)	254254	542542	425425
(b)	258258	582582	825825
(b)	265265	652652	526526
(a)	357357	573573	735735
(a)	475475	754754	547547
(a)	576576	576576	765765
(a)	587587	875875	758758

Finally, the number of fixed points contributed by the non-identity permutations of the set S is $20 + 24 + 12 + 48 = 104$. Therefore the number of distinct jigsaw hexagons formed with six TGPs is, by Burnside's theorem:
$$(1/6)(4096 + 104) = 700. \qquad \square$$

5.4 Hexagon Tile-Figures: the 14 Inner Hexagons

Any hexagon tile formed from six TGPs has an inner hexagon tile-figure. The diagram below shows two examples, together with the ls-labelling of their radii to the goldpoints on the inner sides. These radial ls-strings (measured from a walk-origin) enable us to classify the species of inner hexagons which occur in the 700 possible hexagon tiles. The example tiles have TGP numbers 182652 and 252658, labelled from their 6-chains. Their corresponding radial ls-strings are sllssl and lslssl respectively.

   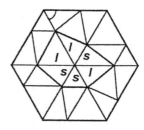

Tile: 182652   **Tile: 252658**

Example jigsaw hexagons, with irregular inner hexagons

Observe that any hexagon tile has an inner hexagon which is well-defined (up to rotations) by its radial ls-string: it is a geometric property of the tile. Any cyclic permutation of the ls-string will represent the same hexagon tile-figure. The numbers of l's and s's in the ls-string are invariant under a cyclic permutation, and their sequential pattern determines the shape of the inner hexagon.

Table 4: Classification of inner hexagon tile-figures

Species (l,s) nos.	subSp.	radial ls–string	freqy. no.	triangles (U, V, W)	symmetries
$A\,(0,6)$		sssss	14	$(6,0,0)$	6 rotation,6 mirror
$B\,(1,5)$		ssssl	64	$(4,0,2)$	1 mirror
$C\,(2,4)$	C_1	ssssll	64	$(3,1,2)$	1 mirror
	C_2	ssslsl	64	$(2,0,4)$	1 mirror
	C_3	sslssl	36	$(2,0,4)$	2 rotation, 2 mirror
$D\,(3,3)$	D_1	ssslll	64	$(2,2,2)$	1 mirror
	D_2	sslsll	64	$(1,1,4)$	none
	D_3	ssllsl	64	$(1,1,4)$	none
	D_4	slslsl	24	$(0,0,6)$	3 rotation, 3 mirror
$E\,(4,2)$	E_1	ssllll	64	$(1,3,2)$	1 mirror
	E_2	slslll	64	$(0,2,4)$	1 mirror
	E_3	sllsll	36	$(0,2,4)$	2 rotation, 2 mirror
$F\,(5,1)$		slllll	64	$(0,4,2)$	1 mirror
$G\,(6,0)$		llllll	14	$(0,6,0)$	6 rotation, 6 mirror
			700		

We have analysed the radial ls-strings of the 700 possible hexagon tiles, and have found that there are just 14 different inner hexagons. The above table summarizes our analysis: and Figure 8 below gives diagrams of all of the inner hexagons. There are 7 species labelled A, B, C, D, E, F, G, and subspecies labelled $C_1, C_2, C_3, D_1, D_2, D_3, D_4, E_1, E_2, E_3$. It will be clear from the representative radial ls-strings why these subdivisions of species occur. The diagrams show how the corresponding tile-figures differ in shape; and dotted lines show axes of symmetry in the inner hexagon figures.

Also given in the table is the distribution of three triangles which occur in various numbers within the interior of each inner hexagon. U refers to an equilateral triangle of side τ^2; V refers to an equilateral triangle of side τ; and W refers to the scalene triangle having sides τ, τ^2, and $\sqrt{2}\tau^{5/2}$.

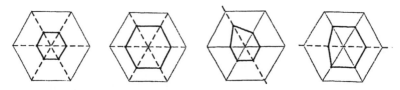

Type A (Ex. 111111)   Type G (Ex. 222222)   Type B (Ex. 111112)   Type F (Ex. 122222)

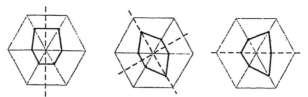

Type C1 (Ex. 111122)   Type C3 (Ex. 112112)   Type C2 (Ex. 111212)

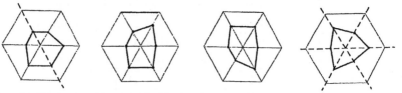

Type D1 (Ex. 111222)   Type D2 (Ex. 112122)   Type D3 (Ex. 112212)   Type D4 (Ex. 121212)

Type E1 (Ex. 112222)   Type E2 (Ex. 121222)   Type E3 (Ex. 122122)

Figure 8. The 14 inner hexagons
(1 ≡ short radius, 2 ≡ long radius)

Chapter 6

Tessellations with Goldpoint Squares

6.1 Introduction

In this Chapter we continue the study of squares marked with goldpoints (goldpoint squares, or SGPs) which was begun in Chapter 4.

First we describe again the six jigsaw-distinct SGPs, adding numbers, colours and ls-strings to their diagrams. Then we use the ls-strings and the group of symmetry rotations of a square to show how Burnside's theorem confirms the existence of six distinct SGPs, and enables the listing of the equivalence classes under the rotations.

After giving the jigging matrix for SGPs, we use it to count jigsaw n-tiles of SGPs, for $n = 2, 3$ and 4. A general solution is given for linear forms.

Finally we show a few tilings of polyhedra, using either all SGPs or else combinations of TGPs and SGPs.

6.2 On the Six Types of Goldpoint Squares

Description of the six types of SGP

In Chapter 4 we gave a diagram showing the six different square tiles that can be obtained by marking a goldpoint into each side of a square. Below, we repeat this diagram, but with much added information. We have numbered them 1 to 6, and assigned the colours Blue (B), Fawn (F), Green (G), Orange (O), Cream (C) and Yellow (Y). We have marked a walk-origin in each left-hand corner (designated a), and we also give the *ls*-string obtained by noting the first steps from the corners when walking anti-clockwise round the square (l = long, and s = short). The corners of squares are always to be imagined with labels a, b, c, d, placed in anti-clockwise order from the walk-origin.

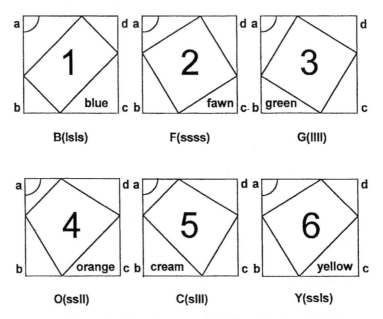

Figure 1. The six types of SGP, with notations

Demonstration that there are only six different SGPs

Before going further with a study of how the SGP types form jigsaws, it is necessary, and instructive, to give a proof that there are only six types of SGP, under jigsaw rules (i.e. allowing for translations and rotations in the plane). So far, we have only given six diagrams and claimed that there are no more types possible.

We begin with the statement that there are $2^4 = 16$ possible ways of marking a square with one goldpoint on each side. This follows from the fact that there are two ways of marking each side, and four sides to mark. Once a square is marked, its *ls*-string can be written down. The list of the 16 possible strings is given in the second column of the table below. They have been labelled 1–to–16, in column one.

Each square may be subjected to four anti-clockwise rotations, of amounts $90°, 180°, 270°, 360°$, the last one returning it to its original position. Using the labels a,b,c,d applied to the square's corners in its original position, we can express these rotations as a group of four permutations, thus:

$$\pi_1 = \begin{pmatrix} abcd \\ bcda \end{pmatrix} \quad \pi_2 = \begin{pmatrix} abcd \\ cdab \end{pmatrix} \quad \pi_3 = \begin{pmatrix} abcd \\ dabc \end{pmatrix} \quad \pi_4 = \begin{pmatrix} abcd \\ abcd \end{pmatrix}$$

Suppose that the permutation π_1 is applied to each of the 16 labelled SGPs in turn. Each will be turned through 90° anti-clockwise, bringing a new corner into the top-left hand position, from which a new ls-string can be recorded. This string will evidently be the original one, but with the last element of it moved up to the front of the string. And of course, the new string must correspond with one which occurred in the original set (say S) of ls-strings. Hence we can say that π_1 operating upon S induces a permutation of the elements of S.

#	ls-string	$\pi_1 S$		$\pi_2 S$	$\pi_3 S$	$\pi_4 S$
1	ssss	ssss	1	1	1	1
2	sssl	lsss	9	5	3	2
3	ssls	sssl	2	9	5	3
4	ssll	lssl	10	13	7	4
5	slss	ssls	3	2	9	5
6	slsl	lsls	11	6	11	6
7	slls	ssll	4	10	13	7
8	slll	lsll	12	14	15	8
9	lsss	slss	5	3	2	9
10	lssl	llss	13	7	4	10
11	lsls	slsl	6	11	6	11
12	lsll	llsl	14	15	8	12
13	llss	slls	7	4	10	13
14	llsl	llls	15	8	12	14
15	llls	slll	8	12	14	15
16	llll	llll	16	16	16	16

Table 1. The four rotation permutations induced on S

In column three of table 1 above, we have shown the results of π_1 operating on each element of S; and in column 4, we have given the original number of each ls-string.

In columns 5, 6 and 7 we have written the element numbers resulting from applications of π_2, π_3, and π_4 in turn to the elements of set S.

Thus columns 2, 4, 5, 6, and 7 give us the four ls-strings for all the 16 SGPs in all their four symmetry rotations. Burnside's Theorem tells us that if we count the number of invariant mappings in all the four induced permutations, and divide by the group cardinal number, we shall obtain the number of equivalence classes of the elements in set S under the group (G) operations.

There are two invariant mappings in columns 4 and 6 (viz. $1 \to 1$ and $16 \to 16$. In column 5, each of 1, 6, 11 and 16 maps to itself. And in column 7, all of the elements map to themselves, since π_4 is the identity element of the group of rotations. Then (using $|G| = 4$):

No. of equivalence classes in $S = (2 + 4 + 2 + 16)/4 = 6$.

This confirms that there are six types of SGP, which are inequivalent under translations and rotations. To describe them, we need only to take a representative from each equivalence class, and note its ls-string. A quick way of doing this is to take the first induced permutation from column 4, and write it out in cycle-form, thus:

$$\pi_1 = \begin{pmatrix} 1 & 2 & 3 & 4 & 5 & 6 & 7 & 8 & 9 & 10 & 11 & 12 & 13 & 14 & 15 & 16 \\ 1 & 9 & 2 & 10 & 3 & 11 & 4 & 12 & 5 & 13 & 6 & 14 & 7 & 15 & 8 & 16 \end{pmatrix}$$

$$= (1)(2, 9, 5, 3)(4, 10, 13, 7)(6, 11)(8, 12, 14, 15)(16)$$

There are six bracketed subsets of S, and these are the type equivalence classes. In the following table, we have chosen a set of representatives which matches with the six diagrams given in Figure 1 above.

lsls (# 11) — rectangle   (fig. 1, number 1)

ssss (# 1)   — first square   (fig. 1, number 2)

llll (# 16) — second square   (fig. 1, number 3)

ssll (# 4)   — trapezium   (fig. 1, number 4)

slll (# 8)   — first quadrangle   (fig. 1, number 5)

ssls (# 3)   — second quadrangle (fig. 1, number 6)

Table 2. Set of representatives for the six SGP types

Thus, using *ls*-strings we have proved that there are six types of SGPs, under jigsaw rules; and we have identified their equivalence classes amongst the 16 fixed aspect SGPs.

6.3 The Jigging Matrix for SGPs

In order to study tiling problems with SGPs, we shall follow procedures similar to those used for TGPs. For example, we shall speak of *n-chains of SGPs*, to mean a horizontal row of SGPs, and they will jig together if the pairs of vertical adjacent sides have complementary *ls*-values.

There are a total of 16 different sub-types of SGP under rotations of the six basic types shown above. We have demonstrated that they are counted as follows: Blue(1) has 2 types; Fawn(2) and Green(3) have 1 type each; Orange(4), Cream(5) and Yellow(6) have 4 types each. The letters at the column heads B, F, G, O, C, Y refer to the six colour types; and the small letters a, b, c, d refer to the sub-types within a colour. Small letters also code the anti-clockwise rotations from the original position, namely $a(0°), b(90°), c(180°), d(270°)$.

We begin developing our jigsaw counting techniques by setting out the possible ways of jigging two SGP's side-by-side, placing the information in a 16×16 jigging matrix. We shall use the same symbol for this (namely M or M^+) as we did for the TGPs matrix. No confusion will arise.

A glance at the jigging matrix M below shows that there are only two different row-patterns of 1s and 0s; and similarly there are only two column-patterns. This seems surprising at first, but once the meaning of the 1s and 0s is understood, the reasons for there being only two row- and two column-patterns become obvious. Theorems 1 to 6 below give the explanations.

	SGP	B_a	B_b	F	G	O_a	O_a	O_a	O_a	C_a	C_b	C_c	C_d	Y_a	Y_b	Y_c	Y_d
	number	1a	1b	2	3	4a	4b	4c	4d	5a	5b	5c	5d	6a	6b	6c	6d
	ls – string	lsls	slsl	ssss	llll	lsls	slsl	ssss	llll	slll	lsll	llsl	llls	ssls	sssl	lsss	slss
B_a	lsls	0	1	1	0	1	0	0	1	1	0	0	0	1	1	0	1
B_b	slsl	1	0	0	1	0	1	1	0	0	1	1	1	0	0	1	0
F	ssss	1	0	0	1	0	1	1	0	0	1	1	1	0	0	1	0
G	llll	0	1	1	0	1	0	0	1	1	0	0	0	1	1	0	1
O_a	ssll	0	1	1	0	1	0	0	1	1	0	0	0	1	1	0	1
O_b	lssl	1	0	0	1	0	1	1	0	0	1	1	1	0	0	1	0
O_c	llss	1	0	0	1	0	1	1	0	0	1	1	1	0	0	1	0
O_d	slls	0	1	1	0	1	0	0	1	1	0	0	0	1	1	0	1
C_a	slll	0	1	1	0	1	0	0	1	1	0	0	0	1	1	0	1
C_b	lsll	0	1	1	0	1	0	0	1	1	0	0	0	1	1	0	1
C_c	llsl	1	0	0	1	0	1	1	0	0	1	1	1	0	0	1	0
C_d	llls	0	1	1	0	1	0	0	1	1	0	0	0	1	1	0	1
Y_a	ssls	0	1	1	0	1	0	0	1	1	0	0	0	1	1	0	1
Y_b	sssl	1	0	0	1	0	1	1	0	0	1	1	1	0	0	1	0
Y_c	lsss	1	0	0	1	0	1	1	0	0	1	1	1	0	0	1	0
Y_d	slss	1	0	0	1	0	1	1	0	0	1	1	1	0	0	1	0

(a) The jigging matrix, with added information (M)

	B	F	G	O	C	Y
B	2	1	1	4	4	4
F	1	0	1	2	3	1
G	1	1	0	2	1	3
O	4	2	2	8	8	8
C	4	3	1	8	6	10
Y	4	1	3	8	10	6

(b) Table 3. The blocked 2-SGP jigsaw counts

Figure 2. Counting 2-SGP jigsaws

Let $S = \{S_i : S_i$ is an SGP, with $i \in 1, 2, ..., 16\}$. And recall that a 1-entry in position (i, j) of M indicates that S_i and S_j can jig as $S_i * S_j$. Otherwise, a 0-entry occurs in that position.

Theorem 1: Two SGPs will jig together if the third side of the left SGP has ls-value which is complementary to the ls-value of the first side of the right SGP.

Proof: By theorems 1 and 2 of chapter 5. $\square$

Theorem 2: The ls-value in the third side of S_i determines the $(0,1)$-pattern in the ith row of M

Proof: The 3rd element of the ls-string of S_i is known, and it is compared with the first element of each of the 16 SGPs in S, taken in turn. These comparisons determine the $(0,1)$-pattern of the ith row of M. $\square$

Theorem 3: There are only two different $(0,1)$ row-patterns in M: and only two different column-patterns in M. In both cases, the two $(0,1)$-patterns are complementary.

Proof: The 3rd ls-value in the ls-string of any SGP is either l or s. Hence, by theorem 1, there exist only two kinds of $(0,1)$ row-pattern in M. A similar argument holds for the columns. It is obvious that the two patterns must be complementary in their 0s and 1s. $\square$

Theorem 4: There are eight 1s and eight 0s in every row and column of M.

Proof: The ls-strings of the 16 SGPs constitute every possible arrangement of an l or an s on each of their four sides. Hence eight of the SGPs will have an l on their 3rd side, and eight a 0. Similarly, eight will have a 1 on their first sides, and eight a 0. The result follows. $\square$

Theorem 5: If the top row of M is filled in (i.e. determined), and the first column is determined too, then the rest of matrix M can be written in directly.

Proof: Obvious, by theorem 3. $\square$

Theorem 6 (powers of M): For $n \geq 2$, $M^n = 2^{3n-4}U$, where U is a 16×16 matrix with every element unity.

Proof: By direct multiplication we find that $M^2 = 4U$. Repeated multiplication by M gives the theorem. $\square$

6.4   Formation of Jigsaw n-Tiles of SGPs

Counting the 2-tiles

Two tiles can be jigged only into a 1×2 linear form (i.e. a 1×2 rectangle). [$90°$ rotations of these produces an equivalent set of 2×1 rectangles.] The set of possible 2-tiles has cardinal number $2^4 \times 2^3 = 2^7 = 128$ members. This total may also be obtained by counting the 1 elements in the jigging matrix; since there are 8 in each row, and there are 16 rows, it is $8 \times 16 = 128$.

As we did with the golden triangles, we can study and count the 2-tiles by blocking the matrix M according to tile colours. The result of that was shown in Table 3, Figure 2. Since no 2-tile can rotate into itself under a $180°$ rotation, the total number of distinct 2-tiles is half the total number of 1s in M, which is 64. This result is, of course, obtained using Burnside's theorem, by application of the group $G(0°, 180°)$, and calculation $(1/2)(2^7 + 0) = 64$. There are no fixed points under rotation through $180°$ since none of tiles $1(a), 1(b), 2$ and 3 can jig with itself.

Counting the 3-tiles

Three SGP tiles can be jigged into two different shapes, namely an angle and a 1×3 linear form:

(i) The angles
$$\boxed{\begin{array}{cc} A & \\ B & C \end{array}}$$
have only 1-fold rotational symmetry. Then, A can be chosen in 2^4 ways, and each of B,C in 2^3 ways. So the total number of distinct angle 3-tiles is $16 \times 8 \times 8 = 2^{10} = 1024$.

(ii) The linear form $\boxed{A|B|C}$ has the same number of possible 3-tiles. But it has 2-fold rotational symmetry, and so fixed points under $180°$ rotations must be counted. There are eight 3-tiles which have this rotational symmetry, namely:

$A\,B\,C$	$A\,B\,C$	$A\,B\,C$	$A\,B\,C$
$1a\ 1b\ 1a$	$1a\ 2\ 1a$	$1b\ 1a\ 1b$	$1b\ 3\ 1b$
$2\ 1a\ 2$	$2\ 3\ 2$	$3\ 1b\ 3$	$3\ 2\ 3$

Hence, by Burnside's theorem, the number of distinct 3-tile linear forms is: $(1/2)(1024 + 8) = 516$.

Counting the linear forms

Using the arguments given above for counting 1×2 and 1×3 linear forms, we can easily show that:

(i) The number of distinct even linear forms, using $1 \times 2n$ tiles of SGPs is:

$$2^4 \times (2^3)^{(2n-1)} = 2^{(6n+1)} \text{ , for } n \in \mathbf{N} \text{ .}$$

This follows from the fact that no linear form of even length can have 2-fold rotational symmetry, since its central pair cannot have that symmetry.

(ii) The number of distinct odd linear forms, using $1 \times (2n + 1)$ tiles of SGPs is:

$$(1/2)[2^4 \times 2^{6n} + 2^{(n+2)}] = 2^{(n+1)}[2^{(5n+2)} + 1] \text{ , for } n \in \mathbf{N} \text{ .}$$

This follows from Burnside's theorem, using the fact that we can construct $2^{(n+2)}$ linear forms of odd length $(2n + 1)$ which have 2-fold rotational symmetry.

An inductive argument proves the fact just stated:

Thus, when $n = 1$ there are the eight $(= 2^3)$ cases just shown above in the table for linear 3-tiles with 2-fold rotational symmetry.

For a linear 5-tile (i.e. $n = 2$) we can begin with any of the 3-tiles from the table, and place it as the central triple: selecting two SGPs (each of the same type) from the type-set $\{1a, 1b, 2, 3\}$ and adding one to each end will complete a linear 5-tile with 2-fold symmetry. An example using this procedure is: $2(1a, 1b, 1a)2$.

This last step can always be carried out in exactly two ways, as we can check from the jigging matrix $\mathbf{M}$. Thus we can construct $2^3 \times 2 = 2^4$ different linear 5-tiles with 2-fold symmetry.

Similarly, we can construct $2^4 \times 2 = 2^5$ linear 7-tiles with that symmetry, by adding an SGP (two choices) to each end of each of the 5-tiles. just constructed. For example, $2(1a, 1b, 1a)2$ can become either $3[2(1a, 1b, 1a)2]3$ or $1a[2(1a, 1b, 1a)2]1a$.

Clearly the number of linear $(2n + 3)$-tiles is twice the number of linear $(2n + 1)$-tiles, by the same argument. And this general inductive step establishes proof of the fact claimed at the outset. $\qquad \square$

Counting the 4-tiles

Four SGP tiles can be formed into seven inequivalent jigsaw 4-tiles. The different possible shapes are shown below, in Figure 3.

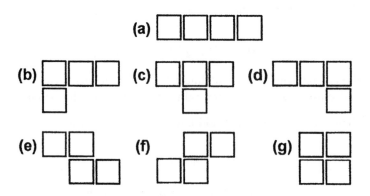

Figure 3. The seven possible 4-tile shapes

The linear form (a) has already been counted, above.

Each of the shapes (b), (c), (d), (e) and (f) may be filled in $2^4 \times (2^3)^3 = 2^{13}$ ways, using SGPs chosen from the 16 available ones (16 for the first tile, then 8 for each subsequent one).

Now, shapes (b), (c) and (d) have 1-fold symmetry, whereas (e) and (f) have 2-fold symmetry; and the square shape (g) has 4-fold rotational symmetry. So we have to act with groups of orders 1, 2 and 4 respectively, on the sets of 4-tiles for each case. The counts for the first five shapes are as follows:

(i)   (b), (c) and (d): There are 2^{13} jigsaw 4-tiles of each of these shapes.

(ii)   (e) and (f): There are $(1/2)(2^{13} + 0) = 2^{12}$ jigsaw 4-tiles of each of these shapes; none provides a fixed point under 180° rotation.

(iii) The 2×2 square (g): This shape may be filled in the following ways: Choose any tile (from 16) to fill the top-left corner: then fill each of the top-right and bottom-left corners (8 available for each): finally, the bottom-right corner can be filled in 2×2 ways, since two of its edges are already determined. The total in the set S of possible such tiles is therefore $16 \times 8 \times 8 \times 4 = 2^{12} = 4096$.

A square has the 4-fold group of rotational symmetries, $G(0°, 90°, 180°, 270°)$. We must consider the action of this group on the set S, and count all the fixed points.

[Four examples of these squares, three having 4-fold symmetry, may be found in Figure 7 of Chapter 4, section 4.62.]

In the following table, we list those combinations of SGPs which will form 2 × 2-tiles with either 2-fold or 4-fold symmetry. Each of the former contributes 1, and each of the latter contributes 3, to the count of fixed points under rotations other than the identity.

$$\textbf{2-fold}: \quad 4^4(2) \quad 1^2 4^2(4) \quad 1^2 5^2(4) \quad 1^2 6^2(4)$$
$$2^2 3^2(2) \quad 2^2 4^2(2) \quad 2^2 5^2(4) \quad 3^2 4^2(2)$$
$$3^2 6^2(4) \quad 4^2 5^2(4) \quad 4^2 6^2(4) \quad 5^2 6^2(4)$$

$$\textbf{4-fold}: \quad 1^4(2) \quad 4^4(2) \quad 5^4(2) \quad 6^4(2)$$

N.B. In the above table, the notation $M^i N^j(k)$ means: i SGPs labelled M can be combined with j SGPs labelled N in k ways, with the stated rotational symmetries.

We can total the fixed points from the above table, and apply Burnside's theorem, thus: the total number of distinct 2 × 2-squares of SGPs is:

$$(1/4)(4096 + 4 \times 2 + 8 \times 4 + 8 \times 3) = 1040.$$

We shall leave problems of counting n-tiles now, although there are clearly many more which could be treated in similar manner to those dealt with above. There are also many challenging problems about tiling the plane, using different types of TGP and/or SGP, which can be posed and attempted. To mention only one, a method is required to determine which of the hexagons of TGPs will tile the plane, and to study the tile-figure patterns engendered by those tilings.

6.5 Tiling of Polyhedra

Introduction

In this final section, we will show how a variety of tiling problems can be solved when the goal is to use TGPs, or SGPs, or both, to tile the sides of given polyhedra.

In particular, we shall give the complete solution for the construction of tetrahedra using six TGPs. Then we shall discuss the tiling of cubes which have sides using one SGP, or four SGPs, or generally, using n^3 SGPs.

Finally, we shall give several diagrams showing how more exotic polyhedra can be tiled; one will use both TGPs and SGPs for its basic tiles.

Producing and counting the TGP tetrahedra

A tetrahedron can be made from four TGPs, as the following diagram shows. Start with a jigging 3-chain A,B,C; then add the fourth TGP labelled D, and fold A and C backwards. Finally D drops on top.

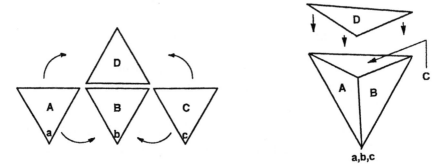

Figure 4. Constructing a tetrahedron

To be able to bring vertices a,b,c together, and to swing C round the back to join A, it is necessary for the first side of A to have the same *ls*-value as the complement of the *ls*-value of the second side of C. To count the 3-chains which have this property, we make use of the following theorem.

> **Theorem 7:** The set (say $J4$) of jigging 4-chains of TGPs which have the same first and fourth TGPs maps one-to-one with the elements of the set (say $J3$) of 3-chains which have the property that the first side of their first TGP will jig to the second side of their third (last) TGP.
>
> **Proof:** If $V \in J4$, then the first three TGPs of U form a 3-chain which belongs to $J3$. We map V to this 3-chain. Conversely, if $W \in J3$, then $W * A$, where A is the first TGP of W, is a jigging 4-chain belonging to $J4$. Since both $J3$ and $J4$ are finite sets, we have established a one-to one mapping of $J3$ to $J4$. (Therefore the two sets have the same cardinal number.) □

We can count $J4$ by adding up the elements on the leading diagonal of the matrix $M^3 = 8U$ This count is $8 \times 8 = 64 = 2^6$. And so, by the theorem just proved, set $J3$ also has 2^6 elements.

To count the tetrahedra, we first note that the side D is uniquely determined by the ls-string, xyz, on the top edges of sides A, B, C. There is one and one only TGP with ls-string $\overline{yzx}$.

Hence there are $2^6 \times 1$ tetrahedra produced in this way. However, they are not all rotationally different; for A, B, C may be rotated about a vertical axis through the common point a, b, c. We have to go back to the 3-chains, and remove rotational redundancies. There are four 3-chains which produce only one tetrahedron, namely those for which $A = B = C$. These cases are from the 3-chains 333, 444, 666, and 888. Each other 3-chain in $J3$ is rotationally equivalent to two others: e.g. $132 \equiv 213 \equiv 321$. Hence the number of possible tetrahedra is $4 + 60/3 = 24$. Below we list the codes for these 24 tetrahedra. The D triangle TGP is given in brackets, after the 3-chain numbers. Two example diagrams are given below the table.

123(7′)	126(3′)	142(6′)	147(2)	173(2)	176(6′)
182(3′)	187(7′)	235(5′)	254(4′)	258(1)	265(1)
333(2)	336(6′)	357(7′)	366(3′)	444(2)	448(6′)
475(6′)	488(3′)	576(5′)	587(4′)	666(1)	888(1)

Table 6. The 24 possible tetrahedra

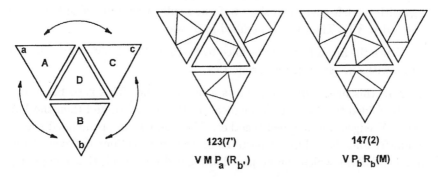

123(7′)

V M P_a ($R_{b'}$)

147(2)

V P_b R_b(M)

Figure 5. Two examples of tetrahedra tilings

Thus we have given a complete solution to the problem of constructing and counting the tetrahedra formed by jigging four TGP tiles.

6.6   On Cubes Tiled with Six SGPs

We have not done a full analysis of this substantial problem: a few partial results will be presented.

How many cubes can be formed?

The answer to this question is easy if rotations in space are not allowed. The counting method proceeds thus: consider the following diagram, of four SGPs in linear form.

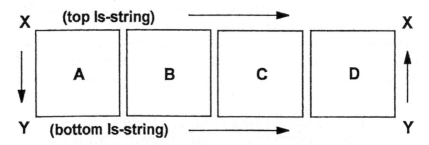

Since there are 16 inequivalent SGPs, if rotations are not allowed, then the cell A can be filled in $16 = 2^4$ ways. Then cell B can be filled in 2^3 ways, if $A * B$ jigs; similarly cell C can be filled in 2^3 ways. Finally, we require D to be filled with an SGP which jigs with C, and whose right-edge jigs with the left-edge of A, since to form a cube we require edge YX to jig with edge XY. Then D can be filled in 2^2 ways.

Hence the linear form can be filled by SGPs in $2^4.2^3.2^3.2^2 = 2^{12}$ different ways, with the given jigging properties.

Now this linear form determines a unique cube. This is so because it can be folded through a right-angle at all the cell joins, and then the YX and XY edges can be glued together. The result is an open-ended square cylinder: and the goldpoint patterns on the open sections uniquely determine two SGPs which can be jigged and glued into them, thus completing a cube.

We conclude that there is a set S of $2^{12} = 4096$ different cubes which can be tiled from the 16 SGPs, provided that rotations in space are not allowed.

To determine how many different cubes there are when rotations are allowed, we may use Burnside's theorem, applying the elements of the cube's symmetry group to each of the cubes in S, and counting fixed points of the induced permutations on S. This complex task remains to be completed.

A lower bound on the number follows immediately from the fact that the rotations group of the cube has order 24, thus:
Let N_c be the number of distinct tiled cubes. Then

$$N_c = \frac{1}{24}(4096 + FP) \geq \frac{4096}{24} > 170 \,.$$

where FP is the total number of fixed points occurring in the 23 permutations induced on S by actions of elements of the symmetry group other than the identity.

The ls-strings of SGPs can be put to good use when studying tiling of cubes, as the following theorem and examples demonstrate.

Some examples of tiled cubes

Theorem 8: (one tile-type only)

A cube tiled with six SGPs each of the same type is possible only with the rectangle tile (type-1) and the trapezium tile (type-4).

Proof:

(i) Type-1: Below we give the developed version of a tiled cube using only type-1 tiles, four in position 1b and two in position 1a. We also give the ls-strings of these two rotation variants of a type-1 SGP.

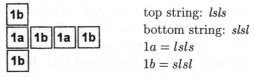

top string: $lsls$
bottom string: $slsl$
$1a = lsls$
$1b = slsl$

Now $1a * 1b$ jigs, since the 3rd side of $1a$ is $l = \bar{s} =$ the complement of first side of $1b$. Similarly $1b * 1a$ jigs. The top and bottom ls-strings of the four tiles forming the central linear form are respectively $lsls$ and $slsl$ when read from left-to-right: hence a tile of type $1b$ will fit onto both the top and bottom open-sections of the folded-up square cylinder.

(ii) Type-4: Below we give the developed version of a tiled cube using only type-4 tiles, three in position 4c, two in position 4b, and one in position 4d. We also give the ls-strings of these rotation variants of a type-4 SGP.

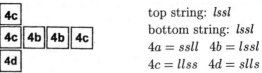

top string: $lssl$

bottom string: $lssl$

$4a = ssll$ $4b = lssl$

$4c = llss$ $4d = slls$

The ls-strings on top and bottom of the linear form are respectively $lssl$ and $lssl$: one can obtain these (top) from complements of fourth sides of the individual tiles, and (bottom) second sides of individual tiles.

Then $4c$ will fit on top of the square cylinder, since its sides (starting with the second one) give $lssl$, as it 'rolls clockwise along the top of the linear form'. And $4d$ will fit on the bottom, since its sides (starting with the fourth, reading its string in reverse, and taking complements) is also $lssl$. Reading the string complement in reverse is because the $4d$ tile in effect 'rolls anti-clockwise along the bottom of the linear form'.

(iii) Types-2,3,5,6: None of these tiles individually will form a cube. To see this, first consider types-2 and -3. They cannot tile a cube because neither of $2 * 2$ and $3 * 3$ jigs; that disposes of them.

In the case of type-5 tiles, one way to form a linear form with four of them is $[5a|5a|5a|5a]$. But with this form, the top and bottom ls-strings are respectively $ssss$ and $llll$, which requires squares of type-2 and type-3 respectively to complete the cube. Other forms of linear forms with four type-5 tiles fail in either this way, or else they do not have the right-end side jigging to the left-end side. Hence six type-5 SGPs cannot form a cube.

Similar arguments hold to show that six type-6 SGPs cannot form a cube.   □

A cube using two types of SGP

The right- and left-ends of the linear form $[4c|1a|4a|1b]$ jig. And the top and bottom *ls*-strings are *llss* and *lssl* respectively. Hence two type-4 tiles (both 4d) will fill the open-sections of the rolled up form, and complete a cube. This example is therefore one of four type-4 tiles and two type-1 tiles forming a cube.

Three cubes using three types of SGP

(i) The right- and left-ends of the linear form [1a|6a|1b|6c] jig. And the top and bottom *ls*-strings are *llsl* and *ssls* respectively. Hence two type-5 tiles (5d on top, 5b underneath) will fill the open-sections of the rolled up form, and complete a cube. This example is therefore one of two type-1 tiles, two type-5 tiles and two type-6 tiles forming a cube.

(ii) The following two examples are of cubes formed from two type-1 tiles, two type-2 tiles and two type-3 tiles. They are different cubes in our set *S*; however, if rotations are allowed, it may easily be seen how one can be rotated into the other. So they are not distinct jigsaw cubes in 3-space.

6.7   Filling Cubes with SGP Cubes

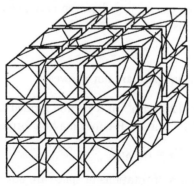

Figure 6. Jigging SGP-cubes

We have studied problems on the formation of $n \times n$ squares, and of $n \times n \times n$ cubes, with unit SGP-tiled cubes. But we have only got complete answers for cases when rotations of the objects are not allowed. We present here four of the basic combinatoric results, whose proofs are very straight-forward.

The first one gives the numbers of ways of jigging unit cubes together with others.

> **Theorem 9:** When building a solid with unit SGP cubes (see Figure 6), any cube placed into the solid must jig with all those cubes which adjoin it (any pair of adjoining sides must have mirror-image patterns). The numbers of ways of adjoining a unit cube to m surrounding ones are:

$$m = 0 : \quad 2^{12} \text{ (number of ways of forming an SGP cube)}$$
$$m = 1 : \quad 2^8$$
$$m = 2 : \quad 2^4 \text{ or } 2^5 \text{ (see the figure above)}$$
$$m = 3 : \quad 2^2 \text{ or } 2^3 \text{ (see the figure above)}$$
$$m = 4 : \quad 2^1$$
$$m = 5 \text{ or } 6 : \quad 2^0$$

> **Theorem 10:** An $n \times n$ square formed from members of the set of 2^{12} unit SGP-cubes may be formed in

$$N_1 = 2^{(5n^2 + 6n + 1)}$$

different ways, if rotations in space are not allowed.

> **Theorem 11:** The number of different $n \times n$ squares of SGP cubes which can be jigged on top of a given $n \times n$ square of SGP-cubes is

$$N_2 = 2^{(3n^2 + 4n + 1)}$$

if rotations in space are not allowed.

> **Theorem 12:** An $n \times n \times n$ cube formed from members of the set of 2^{12} unit SGP-cubes may be formed in

$$N_3 = N_1 \times N_2^{n-1} = 2^{[3n(n+1)^2]}$$

different ways, if rotations in space are not allowed.

6.8   A Variety of Solids Tiled with TGPs and SGPs

To end this chapter we give seven diagrams which display exploded plans of solids which are tiled with TGPs (equilateral triangles of side 1, with goldpoints) and SGPs (squares of side 1, with goldpoints).

Included are a truncated tetrahedron, two tilings of a solid known as a rhombocubooctahedron, two tilings of a cubooctahedron, an 80-sided TGP-tiled polyhedron, and Kepler's stella octangula. For details of these solids, consult the book *Polyhedron Models* by M. Wenninger, Cambridge University Press (1971).

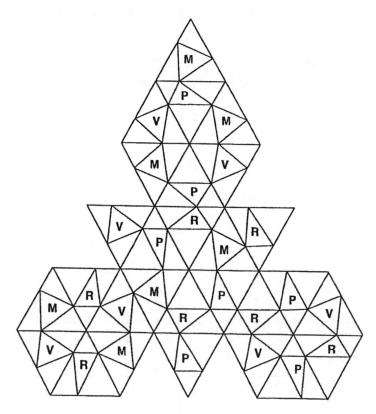

A truncated tetrahedron

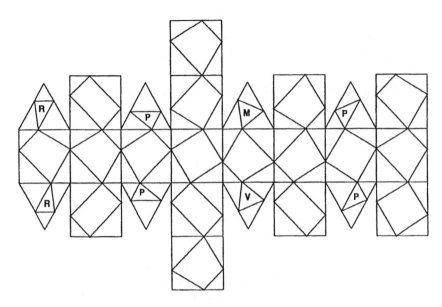

First rhombocubooctahedron

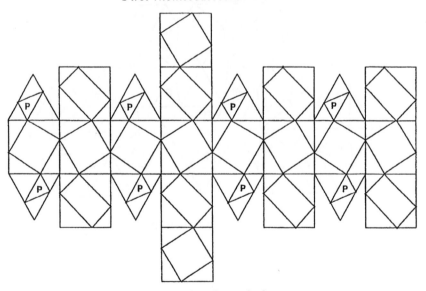

Second rhombocubooctahedron

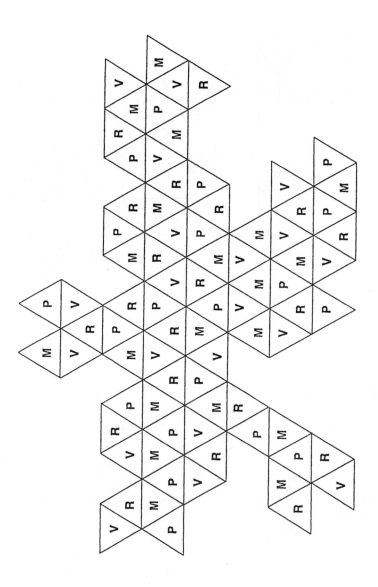

An 80-sided polyhedron; no neighbouring sides of same colour

*(goldpoints not shown: they may easily be added,
working from the V- and M-tile markings)*

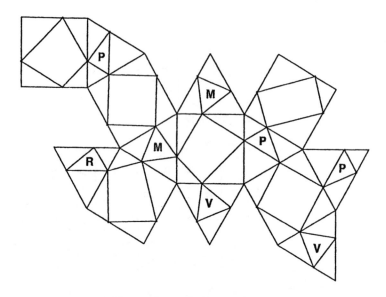

First cubooctahedron

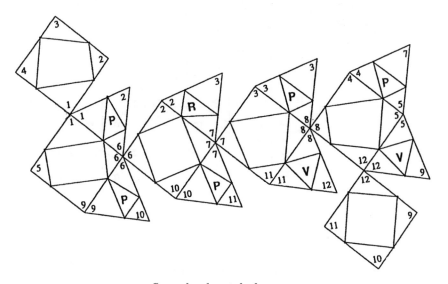

Second cubooctahedron

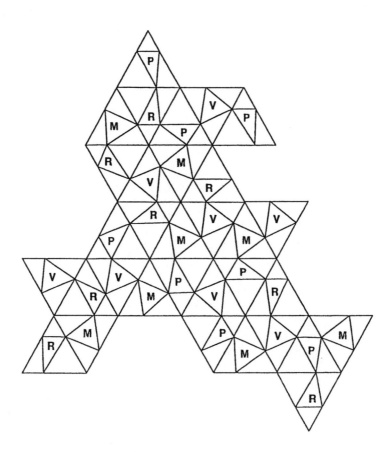

Kepler's star-like octahedron (stella octangula)

Chapter 7

Games with Goldpoint Tiles

7.1 Introduction

It is possible to design interesting games which use the goldpoint triangles (TGPs) and goldpoint squares (SGPs) as pieces. In this chapter we describe the rules for playing several types of game, each of which exploits the variety of the TGP and SGP tiles and their tile figures. The moves which the players can make exploit the jigging operation. Indeed, the games might be called jigsaw games using golden tiles as jigsaw pieces. Anyone who enjoys doing jigsaw puzzles, and is a mathematician with a taste for geometric figures and combinatorics, will enjoy playing these games. Problems for mathematical study within goldpoint geometry will doubtless suggest themselves, during play.

Some of the games can be played by one person *à la solitaire*, whilst others are competitive, and can be played by two or more players.

7.2 Equipment Needed for the Games

First it is necessary to prepare a sufficient quantity of the TGPs and SGPs as described in chapters 4 and 5. For any particular game, a collection of these tiles (or *pieces* for a jigsaw game) will be prescribed, and that will be known as the *pool* for the game. Players will be required to select a given number of pieces from that pool, before starting the game. A player's selection from the pool will be called his or her *hand*, as in card games.

Constructing the pieces is a relatively simple matter, especially if one has the use of a computer-aided drawing package. It is recommended that equilateral triangles and squares be drawn with side length 5 cm., and the goldpoints be marked approximately, say at a 2 cm. or a 3 cm. distance from each corner (1.9 cm. and 3.1 cm. are better approximations). Then the tile figures can be drawn, joining up the goldpoints in sequence around the triangle or square. Consult the diagrams in chapter 5 to get all the details correct. One corner (the 'a' corner) on each tile should be angle-marked; this makes it very easy to determine which sub-type (rotant) * a tile is, according to its position relative to the horizontal at any time. Finally, the number of each tile type must be printed, in large font-size, in the centre of the tile The numbers used must be 1–4 for the TGPs and 1–6 for the SGPs, assigned in the orders of their colours, being respectively V, M, P, R for the TGPs and B, F, G, O, C, Y for the SGPs.

It is only necessary to draw one example of each type of TGP and SGP. Then, using copying techniques, these can be multiplied to fill an A4 sheet. About twenty of each can be so arranged. Four sheets of TGPs and six sheets of SGPs should be prepared like this.

Finally, take the ten prepared sheets of diagrams to a copy-shop, and have them copied onto thin cardboard sheets, each in the right colour, as specified above. That is: violet, mauve, pink, red for the TGPs; and blue, fawn, green, orange, cream, yellow for the SGPs. It would be wise to have two card copies made of each, and to keep the spares and master sheets for future use.

Have the copy-shop guillotine the pieces from one card of each type (or cut them out carefully with scissors). Then you will have an overall pool of 200 pieces, twenty of each type of golden tile. This will be adequate for all the games to be described below.

With some games it will be recommended that a *board* be drawn up, on which to place and move the pieces around. We shall give details of those when they are required. In most games, the playing surface will be a smooth table top. A large wooden tray might make a good surface, also keeping the pieces together and enabling games in play to be portable.

---

*We define *rotant* to mean a 'rotational variant' of a tile. The TGP types 1 and 2 have only one distinct rotant each, whereas TGP types 3 and 4 have three distinct rotants each.

We next describe several games which can be played with TGPs, using only the four types (all the rotants), allowing for translations, and rotations of the equilateral triangles through integer multiples of 60°.

7.3 Games with Goldpoint Triangles (TGPs)

7.31 GOLDPOINT TRIANGLE DOMINOES

THE POOL: Twelve TGPs of each colour are mixed well, then formed into a stack, which is placed in the middle of the table.

THE HANDS: Each player (2–4 players) draws a hand of six pieces, taking them one at a time in turn from the top of the stack.

THE OBJECTIVE: The aim of the game is to get rid of all one's pieces, by placing them on the table in the manner specified by the rules given below.

THE PLAY: Like in dominoes, a line (linear form) is to be formed with the pieces. The first player is chosen by tossing a coin, or drawing cards from a pack. The first player places a piece on the table. The second player (on his left) must place a piece alongside the first, jigging them, and thereby determining the line orientation.

The third player (to the left) must extend the line, in either direction, by jigging one of her pieces onto an end tile. And so on, as each player plays in turn, anti-clockwise round the players.

If a player is unable to place a piece on the line, when taking a turn, then she forfeits the turn and must pick up a piece from the top of the stack, and add it to her hand.

If a line reaches an edge of the table, it is slid lengthwise to the middle of the table. Once a line can be extended no more, the next player in turn begins a new line.

Play continues thus until a player uses the last piece from her hand, and she is declared the winner.

7.32 GOLDPOINT HEX SOLITAIRE:

THE BOARD: For this game, it is helpful to have a board (white cardboard) on which is drawn a regular hexagon, of side length just slightly longer than 5cm, with the six radii drawn in to its centre.

THE POOL: Six TGPs of each type are shuffled together, to form a 36-piece stack.

THE GAME: The player draws six tiles from the top of the stack, and tries to arrange them into the hexagon on the board, jigging all pieces correctly. If she fails to do so, she draws a further three tiles from the stack; and then tries to complete the hexagon WITHOUT REMOVING ANY TILES ALREADY PLACED, AND WITH NO FURTHER REARRANGEMENTS OF THOSE TILES BEING ALLOWED. If she succeeds, then that is a winning game. If not, the game is abandoned, and another begun.

VARIATION: This time, the objective is to complete hexagons and to maximize a score which is determined by the kinds of inner hexagons (tile figures, see Ch. 5, Fig. 8) achieved. The following scoring table is suggested.

Inner hexagon:	A	B	C	D	E	F	G
Score:	15	6	3	2	3	6	15

The player proceeds as above, but at any one turn he may decide not to use a piece which will jig into the hexagon, and instead draw the next three tiles from the stack. This option may only be used once in the course of the play. It might help achieve a higher scoring inner hexagon.

A player cannot start another hexagon until the current one is completed. The game is repeated until five hexagons have been completed, and then the total score for the five inner hexagons is the player's final score.

COMBAT VARIATION: If two players each form five hexagons, separately with two stacks, and total the scores for their five inner hexagons, the winner will be the one achieving the highest score. They will play to a fixed time limit (e.g. 15 minutes), and their total score will be computed from their completed hexagons.

7.33 GOLDPOINT HEX COMBAT (two players)

THE BOARD: As for Goldpoint Hex Solitaire.

THE POOL AND HANDS: As for the dominoes game.

THE PLAY: The starting player is decided by coin toss. He places a piece inside the hexagon, anywhere. The next player does likewise; if her piece goes alongside the first one, it must jig; but it need not be adjacent to the first.

Play continues thus until a player cannot place a piece. He or she then forfeits the turn, and takes another piece from the stack pool.

THE OBJECTIVE: The aim is to complete a hexagon jigsaw; and the first player to fill the hexagon on the board is declared the winner.

VARIATION: When a player succeeds in placing a piece, he is allowed a second turn.

7.4 Games with goldpoint squares (SGPs)

7.41 NOUGHTS AND CROSSES

This simple game, beloved by all youngsters, takes on new and interesting aspects when played with goldpoint squares. We offer three versions, but more are possible. Players can invent variations for themselves. It is a game for two players, which we shall call A and B.

THE START: A tosses a coin and B calls 'head' or 'tails'. The winner of the toss starts the game. Suppose it is A.

THE HANDS: A takes 5 pieces (i.e. SGPs) of any one colour she wishes. Then B takes 5 pieces of any other colour.

THE PLAY: The play proceeds by players taking turns (A starts) in putting down a piece on the table, in order to fill a 3×3 square. Whenever a piece is placed by the side of one or other previously placed pieces, it must jig there properly. If a player is unable to place a piece, on her turn, then she forfeits that turn.

THE OBJECTIVE: The first player to get a full line of pieces of her own colour, either row, column or diagonal, is the winner.

THE WIN OR DRAW: The play proceeds until one player is declared winner, or else a position is reached whereby both players are unable to place another piece, in which case a draw is declared.

VARIATIONS:
(1) Colour choice is important! The two colours to be used may be fixed beforehand.
(2) Player A chooses 3 pieces of one colour and 2 of another colour; then B chooses 3 and 2 pieces of two colours different from A's. Then the winner is any player who first places a line of any three pieces from his hand.

(3) Draw hands of 8 pieces, and play to fill a 4 × 4 square. Win with a line of any three pieces (or 4 pieces – decided beforehand).

7.42 GOLDPOINT SQUARE DOMINOES

THE POOL: Twelve SGPs of each colour are mixed well, then formed into a stack, which is placed in the middle of the table.

THE HANDS: Each player (2–4 players) draws a hand of six pieces, taking them one at a time in turn from the top of the stack.

THE OBJECTIVE: The aim of the game is to get rid of all ones pieces, by placing them on the table in the manner specified by the rules given below.

THE PLAY: Like in dominoes, a line (linear form) is to be formed with the pieces. The first player is chosen by tossing a coin, or drawing cards from a pack. The first player places a piece on the table. The second player (on his left) must place a piece alongside the first, jigging them, and thereby determining the line orientation.

The third player (to the left) must extend the line, in either direction, by jigging one of her pieces onto an end side. And so on, as each player plays in turn, anti-clockwise round the players.

If a player is unable to place a piece on the line, when taking a turn, then she forfeits the turn and must pick up a piece from the top of the stack, and add it to her hand.

If a line reaches and edge of the table, it is slid lengthwise to the middle of the table. Once a line can be extended no more, the next player in turn begins a new line.

Play continues thus until a player uses the last piece from her hand, and she is declared the winner.

7.43 FILLING THE SQUARES

SOLITAIRE (1): From a well-shuffled stack of 72 pieces (12 of each colour), the player draws 13 pieces from the top. The objective is to form a 3 × 3 square and a 2 × 2 square with the 13 tiles, jigging them properly together.

If this objective is not achieved, the player draws a further 3 pieces from the stack, and tries to complete the two squares, without rearranging them. If this attempt fails, the game is abandoned, and the 16 pieces are shuffled into the stack, to begin a new game.

SOLITAIRE (2): As for (1), but drawing 25 pieces from the stack, and trying to form a 4 × 4 square and a 3 × 3 square.

SOLITAIRE (3): Have an unlimited supply of SGPs, and try to fulfil certain conditions or challenges on the $n \times n$ tile to be produced. Example challenges are:
(1) Produce an $n \times n$-tile that will tile the plane (a) with one type of SGP, (b) with two types of SGP, (c) with 3 types of SGP, and so on.
(2) Produce a Latin square (a) of n differently coloured tiles, (b) of n different tile-types (i.e. rotants). (3) Produce a Graeco-Latin square with a specified set of tiles (e.g. of specified colours and specified rotants).

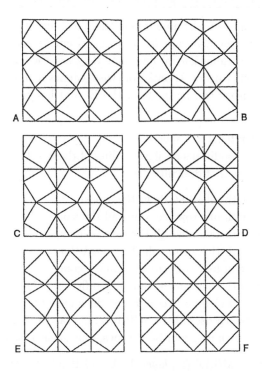

Figure 1. Six tiled 3 × 3 *squares*

In Fig. 1 we show six 3 × 3-tiled squares, to illustrate some possibilities. In the table below, we give their written notations; it is very easy to write these down, row-wise, and thereby keep a record of the tiles you have achieved. After a little time spent on this activity, you will find which are the hardest challenges to meet, and begin to devise strategies for achieving them.

The notations for the squares are:

	row 1			row 2			row 3		
A :	4b	4c	4b	4a	4d	4a	4b	4c	4b
B :	6d	5b	6d	6c	3	6a	6b	5d	6b
C :	3	2	3	2	3	2	3	2	3
D :	6b	6c	5c	5c	3	6a	6d	4	5c
E :	4c	4b	4c	4d	4a	4d	4c	4b	4c
F :	1a	1b	1a	1b	1a	1b	1a	1b	1a

The F-square is made up of nine type-1 tiles (rectangles).

The A-square and the E-square are both constructed with nine type-4 tiles (trapezia). On examination we find that they are the same 3×3 jigsaw square, since a 90° turn of A, anti-clockwise, carries it into E.

Pleasing comparisons can be made of the E- and F-squares, in regard to their tile figures.

Challenges for the 4×4 *square*

The following conditions on jigsaw tiling of the 4×4 square pose problems of varying levels of difficulty.

(1) Solution with all 16 SGPs in the a-position.
(2) Solution with one row (column) all the same colour.
(3) Solution with two rows all the same colour.
(4) Solution with three rows all the same colour.
(5) Solution with four rows all the same colour.
(6) Solution with each row having four different colours.
(7) Solution in which no two adjacent sides have the same colour.
(8) Solution which uses all 16 SGP rotants (1a, 1b, 2, ... , 6b, 6c, 6d).
(9) Solution which uses four SGP-types, four of each, as a Latin Square.

SQUARES COMBAT (2 players): A and B toss for start. Assume A wins. Then A and B in turn draw 4 pieces from the stack, again and again until they each have 16 pieces.

Player A places a piece on the table. B jigs one onto it. And so on, each trying to form a 2×2 square. The first player to do this scores two points. Then the two players continue, adding pieces in turn, in order to build the

2 × 2 square into a 3 × 3 square. The first player to complete this scores 3 points.

This process continues until a 4 × 4 square is built. The first player to complete this scores 4 points.

This whole process is now restarted, with the player who DID NOT score the last 4 points placing the first piece.

If a position is reached such that a player cannot place a piece, he forfeits that turn.

If a position is reached such that neither player can place a piece, then a new 2 × 2 square is started, by the player whose turn it was when play was found to be blocked.

And so on, until all 32 pieces have been placed on the table.

THE WINNER: The player who has scored most points during the play is declared the winner.

7.44 THE GAME "15" (solitaire)

Take a set of the 16 rotants of SGPs. Arrange them as a goldpoint jigsaw on a 4 × 4 chessboard (any rotation of a piece is allowed, so long as it jigs with its adjacent pieces). Now remove any piece (say a central one) leaving a vacant cell. Call the piece P, and lay it aside.

The purpose of the game is to move (slide) a piece into the vacant cell – from above, below, or either side – leaving another vacant cell. Continue this process until (i) a goldpoint jigsaw of the 15 pieces is reached, and (ii) the piece P jigs into the final vacant cell. Thus another goldpoint jigsaw is arrived at, filling the 4 × 4 board.

Of course, other rules can be imposed upon allowable juxtapositions, to make the game more complex. Try introducing one or two.

7.45 CAPTURING 4-SQUARES

EQUIPMENT: An 8 × 8 chess board, with cells to take SGPs.

THE PLAY: Their are two players. Each has a set of the sixteen SGP rotants. One set is all in colour white (say) and the other all in another colour (blue, say).

THE OBJECTIVE: Each player tries to fill as many 2 × 2 squares as he can, with his tiles. Of course, they will try to prevent their opponents from achieving this goal, by using their tiles wisely.

THE PLAY: All tiles are to be kept face-up, and visible to both players. Players place a piece in turn, into a cell on the board. When placing an SGP, it must jig with all surrounding pieces.

The game ends when all tiles are used up, or when neither player can place another tile.

If at any stage a player cannot place a tile, he forfeits that turn.

Completed 2 × 2 squares remain on the board, and players can add tiles around them. If, for example, a player completes a 2 × 4 rectangle, he/she will gain scores for two 2 × 2 completions.

SCORING: When the game ends, each player totals the number of 2 × 2 squares he/she has covered. The winner is the one with the highest total.

7.46 GOLDPOINT CHESS

Chess is one of the oldest, best and most popular two-person games in the world. Over the centuries there have been many versions of this 'battle board-game', but the rules of play are now standard over most countries. However, we now offer a slightly revised version, which does complicate the play, and which requires new strategies to be devised. Perhaps some readers of this book will try out this variety of chess, and judge whether it is worth persevering with.

It uses the square goldpoint tiles for its pieces; and it is played with all the normal moves and rules of standard chess (which we assume are known by the reader). The addition that we make is called *the jigsaw rule*, which must be obeyed whenever a piece is moved into an empty cell; then the piece must jig with all pieces which are already in position around that cell. The only piece which is exempted from the jigsaw rule is the King, which can only move one step anyway, and often has great need to move into an empty cell adjacent to it.

EQUIPMENT: An 8 × 8 chess board is required, whose squares are big enough to comfortably accommodate the SGP chess pieces.

Two sets of 16 chess 'pieces' must be prepared, one set for each player, using two different colours (or shadings) for the sets.

All this equipment can be prepared from diagrams which we present later in figure 2.

Using our notations for SGPs, we can describe the pieces as follows:

Second row:	6a 5a 6a 5a 6a 5a 6a 5a
First row:	4b 1a 4a 2 3 4d 1b 4c

Note that this arrangment is for the (traditionally) white player. So the white Queen is the 2-tile, and the white King is the 3-tile.

When this arrangement is swung around through 180°, it becomes the black player's pieces; the black Queen is now the 3-tile, and the black King is the 2-tile. Otherwise, the corresponding pieces have different colours but the same notations and tile figures. And, of course, First row now means Eighth row, and Second row means Seventh row on the board.

In order that the players know immediately which piece is which, the SGPs must be marked with letters as follows:

P (pawn), R (rook), N (knight), B (bishop), Q (queen), K (king)

It is not necessary to mark the major pieces 'left' or 'right', as the starting positions are already defined, and pieces are allowed to rotate when moving into new cells.

The diagram on page 305 shows a set of pieces, and also gives two rows of a board. Anyone wishing to play this game can copy the diagram (at double size and as many times as needed), and quickly make up a chess set by sticking the copied sheets onto card, and cutting out the pieces. The black pieces have a dark-shaded area on them, to distinguish them from the white pieces.

THE PLAY: Play proceeds exactly as in standard chess, except that the following jigsaw rules must be observed. The objective is the same, namely to checkmate one's opponent's King.

THE JIGSAW RULES: The Kings are exempt from the jigsaw rules. All other pieces must obey them.

(1) When a piece is moved to a cell which is empty, it may be rotated to any position, but when placed in the cell it must jig with all pieces already adjacent to the cell.

(2) When a piece is moved to an occupied cell, for the purpose of capturing an enemy piece, it may take and remove the occupying piece, regardless of whether it can properly jig with all adjacent pieces. Moreover, it can be rotated into any position before being placed in the cell.

(3) When a pawn is taking *en passant*, rule (1) is waived, and the pawn may rotate before placement.

(4) Rule (1) is waived for both pieces taking part in a 'castling move'. The rook may rotate before placement.

FINAL COMMENT: We have made the chess jigsaw rules as simple as possible, so that they will be remembered and applied easily. We hope that readers who try Goldpoint Chess will not feel frustrated by Rule (1), but will enjoy the challenge of finding new strategies, and find pleasure in the ways in which tile figures can be observed and used to help grapple with the added complexities of the game.

7.47 GOLDPOINT RUBIK'S CUBE

Figure 6, Ch. 6, shows a cube made up of 3^3 tiled SGP-cubes. This object immediately suggests that a Rubik's cube could be constructed of SGP-cubes, and the goldpoint tile-figures on the outer sides used to add to the complexities of 'solving' it. We assume that the reader knows what is meant by this — that the pattern on the sides of a Rubik cube can be scrambled by rotating layers of the constituent cubes, and to solve the puzzle means to find a sequence of rotations which will cause the original pattern to be restored.

As is well known, Rubik's cube took the world by storm in the late 1970s, and a huge literature and mathematical theory developed around it. A fascinating account of this explosion of cube activity may be read in *Metamagical Themes: Questing for the Essence of Mind and Pattern*, D. R. Hofstadter, Basic Books Inc., N. Y. [1985, pp. 301–363]. During these times the Rubik's cube captured the imagination and perseverance of millions of people, many of whom became afflicted by the sickness which Hofstadter describes thus, from a medical dictionary entry:

> **Cubitis magikia, n** *A severe mental disorder accompanied by itching of the fingertips, which can be relieved only by prolonged contact with a multicoloured cube originating in Hungary and Japan. Symptoms often last for months. Highly contagious.*

It is evident that Goldpoint Rubik's Cubes, by posing a virtually indefinite list of new patterns and challenges to puzzlers, would greatly add to the risks of contracting that mental disorder.

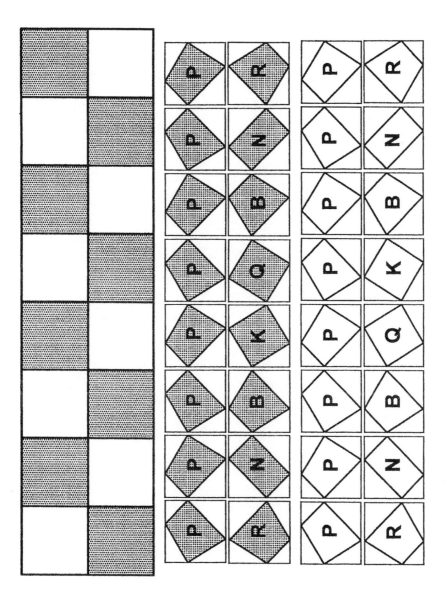

Figure 2. Set of chess pieces, and portion of board

Bibliography

Part B, Section 2

Goldpoint Geometry

1. ATANASSOVA, V. K. and Turner, J. C. 1999: On Triangles and Squares Marked with Goldpoints — Studies of Golden Tiles. In F. T. Howard (Edr.), *Applications of Fibonacci Numbers,* **8**, Kluwer Academic Press, 11–26.
2. DODD, F. W. 1983: *Number Theory in the Quadratic Field with Golden Section Unit.* Polygonal Publishing House, 80 Passaic Ave., Passaic, NJ 07055.
3. HUNTLEY, H. E. 1970: *The Divine Proportion.* Dover Publications Inc.
4. KNOTT, R. 2000: *Web page.*
   http://www.mcs.surrey.ac.uk/R.Knott/Fibonacci/
5. LAUWERIER, H. 1991: *Fractals — Images of Chaos.* Penguin Books.
6. MILNE, J. J. 1911: *Cross-Ratio Geometry.* Cambridge University Press.
7. TURNER, J. C. and Shannon, A. G. 1998: Introduction to a Fibonacci Geometry. *Applications of Fibonacci numbers,* **7**. (Eds. G. E. Bergum et als.) Kluwer A. P., 435–448.
8. WALSER, H. 1996: *Der Goldene Schnitt.* T.V.L. und vdf Hochschulverlag AG an der ETH Zürich.

Index